Synthesis Lectures on Data Mining and Knowledge Discovery

Editors
Jiawei Han, *UIUC*
Lise Getoor, *University of Maryland*
Wei Wang, *University of North Carolina, Chapel Hill*
Johannes Gehrke, *Cornell University*
Robert Grossman, *University of Chicago*

Synthesis Lectures on Data Mining and Knowledge Discovery is edited by Jiawei Han, Lise Getoor, Wei Wang, Johannes Gehrke, and Robert Grossman. The series publishes 50- to 150-page publications on topics pertaining to data mining, web mining, text mining, and knowledge discovery, including tutorials and case studies. The scope will largely follow the purview of premier computer science conferences, such as KDD. Potential topics include, but not limited to, data mining algorithms, innovative data mining applications, data mining systems, mining text, web and semi-structured data, high performance and parallel/distributed data mining, data mining standards, data mining and knowledge discovery framework and process, data mining foundations, mining data streams and sensor data, mining multi-media data, mining social networks and graph data, mining spatial and temporal data, pre-processing and post-processing in data mining, robust and scalable statistical methods, security, privacy, and adversarial data mining, visual data mining, visual analytics, and data visualization.

Graph Mining: Laws, Tools, and Case Studies
D. Chakrabarti and C. Faloutsos
2012

Mining Heterogeneous Information Networks: Principles and Methodologies
Yizhou Sun and Jiawei Han
2012

Privacy in Social Networks
Elena Zheleva, Evimaria Terzi, and Lise Getoor
2012

Graph Mining

Laws, Tools, and Case Studies

Community Detection and Mining in Social Media
Lei Tang and Huan Liu
2010

Ensemble Methods in Data Mining: Improving Accuracy Through Combining Predictions
Giovanni Seni and John F. Elder
2010

Modeling and Data Mining in Blogosphere
Nitin Agarwal and Huan Liu
2009

Graph Mining: Laws, Tools, and Case Studies
D. Chakrabarti and C. Faloutsos

ISBN: 978-3-031-00775-0   paperback
ISBN: 978-3-031-01903-6   ebook

DOI 10.1007/978-3-031-01903-6

A Publication in the Springer series
SYNTHESIS LECTURES ON DATA MINING AND KNOWLEDGE DISCOVERY

Lecture #6
Series Editors: Jiawei Han, *UIUC*
　　　　　　Lise Getoor, *University of Maryland*
　　　　　　Wei Wang, *University of North Carolina, Chapel Hill*
　　　　　　Johannes Gehrke, *Cornell University*
　　　　　　Robert Grossman, *University of Chicago*
Series ISSN
Synthesis Lectures on Data Mining and Knowledge Discovery
Print 2151-0067   Electronic 2151-0075

Graph Mining

Laws, Tools, and Case Studies

D. Chakrabarti
Facebook

C. Faloutsos
CMU

SYNTHESIS LECTURES ON DATA MINING AND KNOWLEDGE DISCOVERY #6

ABSTRACT

What does the Web look like? How can we find patterns, communities, outliers, in a social network? Which are the most central nodes in a network? These are the questions that motivate this work. Networks and graphs appear in many diverse settings, for example in social networks, computer-communication networks (intrusion detection, traffic management), protein-protein interaction networks in biology, document-text bipartite graphs in text retrieval, person-account graphs in financial fraud detection, and others.

In this work, first we list several surprising patterns that real graphs tend to follow. Then we give a detailed list of generators that try to mirror these patterns. Generators are important, because they can help with "what if" scenarios, extrapolations, and anonymization. Then we provide a list of powerful tools for graph analysis, and specifically spectral methods (Singular Value Decomposition (SVD)), tensors, and case studies like the famous "pageRank" algorithm and the "HITS" algorithm for ranking web search results. Finally, we conclude with a survey of tools and observations from related fields like sociology, which provide complementary viewpoints.

KEYWORDS

data mining, social networks, power laws, graph generators, pagerank, singular value decomposition.

Christos Faloutsos: *To Christina, for her patience, support, and down-to-earth questions; to Michalis and Petros, for the '99 paper that started it all.*

Deepayan Chakrabarti: *To Purna and my parents, for their support and help, and for always being there when I needed them.*

Contents

Acknowledgments

Several of the results that we present in this book were possible thanks to research funding from the National Science Foundation (grants IIS-0534205, IIS-0705359, IIS0808661, IIS-1017415), Defense Advanced Research Program Agency (contracts W911NF-09-2-0053, HDTRA1-10-1-0120, W911NF-11-C-0088), Lawrence Livermore National Laboratories (LLNL), IBM, Yahoo, and Google. Special thanks to Yahoo, for allowing access to the *M45* hadoop cluster. Any opinions, findings, and conclusions or recommendations expressed herein are those of the authors, and do not necessarily reflect the views of the National Science Foundation, DARPA, LLNL, or other funding parties.

D. Chakrabarti and C. Faloutsos
September 2012

CHAPTER 1

Introduction

What does the web look like? How about a social network, like LinkedIn or Facebook? Is it normal for a person to have 500 "friends?" Is it normal to contact all 500 with equal frequency? Which are the most important nodes in a graph?

These are some of the motivating questions behind this book. We want to explore what typical ("normal") networks and nodes look like, so that we can spot anomalies (e.g., hi-jacked accounts, phishers, spammers, and mis-configured routers in a computer network). We also want to answer *what-if* scenarios, to do link prediction, and study how viruses and rumors propagate and how to stop them (or accelerate them).

Informally, a graph is set of nodes, pairs of which might be connected by edges. The graph representation is simple, yet powerful, encompassing a large number of applications:

- *social networks*, like Facebook and LinkedIn, which have brought on a revolution in this decade. All contact networks fall under the same class: who-calls-whom in a phone network, who-texts-whom, or who collaborates with whom.
- *cyber-security*, for computer networks, where machines send IP packets to each other. The goals are to spot strange behaviors, like botnets, port-scanners, exfiltrators (e.g., industrial espionage).
- *web and ranking*: the web consists of pages pointing to each other, and successful ranking algorithms must analyze this graph intelligently.
- *influence propagation*: how do new concepts, terms, "memes," propagate? What can we do to accelerate a word-of-mouth campaign for our new product? Conversely, what can we do to stop the propagation of an undesirable element (like the flu, or a computer virus).
- *e-commerce*: In several cases, especially in Web 2.0, users declare what products they have bought or liked. Amazon, Netflix, and the Apple and Android app stores are just a few of the many cases where this information can be used to increase user satisfaction, as well as profits. Similarly, Ebay allows buyers to give ratings to sellers; the shape of the resulting sub-graph could help spot fraud [226].
- *everything else*: in biology, protein-protein interaction networks can be modeled as graphs. In population ecology, we have food-webs of species in prey-predator relationships. In text retrieval, we have bi-partite graphs of documents linked to the terms they contain, in the "vector space" model (also know as the "bag of words" model).

In the rest of this book we present surprising, recurring patterns, among all these diverse settings. One may wonder what the practical use is of such patterns, such as the famous "six degrees of separation," which claims that social contact networks have a short diameter of about 6. We present the definitions and details later (see Chapter 2), but the idea is that most people are about six handshakes away from most other people in the world(!).

There are several reasons to study patterns and laws in real graphs:

- **Understanding** human (machine, protein) behavior: Discovering patterns of connectivity may help domain experts (sociologists, biologists) develop even better theories, explaining how and when humans (machines, proteins) come into contact with others.

- **Anomaly detection**: if we know we should expect a specific pattern (like, say, the "six degrees"), then a graph with an extremely short diameter (say, a "star"), or extremely long diameter (say, a "chain"), would be suspicious. In fact, there is no perfect, formal definition of anomaly, and similarly, there is no perfect definition of "normal:" An anomaly is an event (node, subgraph), whose properties are rare, and thus deviate from the majority of the population. Informally, "normal" behavior is the pattern that the majority follows, and "abnormal" behavior is the deviation. Thus, anomalies and patterns go hand-in-hand. We also want to spot local anomalies, that is, strange subgraphs/edges/nodes. All such abnormalities should deviate from the "normal" patterns, so understanding the patterns that appear in natural graphs is a prerequisite for detection of such outliers.

- **Extrapolations**: What will the web look like next year? If a growing network had a diameter of 6 for all past years, should we expect a slight growth in the diameter for next year? In fact, the diameter exhibits particularly strange trends – see Chapter 3.

- **Generator design**: We list almost a dozen graph generators in Part II – which one(s) are the most realistic? Knowing the behavior of "normal" graphs can help us avoid un-realistic generators. An important need for generators is for simulations: Real graphs often have privacy issues and/or high monetary value, and cannot be released; in that case researchers can try their algorithms on synthetic, but realistic-looking graphs. For example, in order to test the next-generation Internet protocol, we would like to simulate it on a graph that is "similar" to what the Internet will look like a few years into the future.

- **Graph compression**: Graph patterns represent regularities in the data. Such regularities can be used to better compress the data. For Internet-size graphs, like web-crawls with billions of pages, or Facebook with "more than 950 million active users"[1] at the time of writing, keeping (and compressing) historical data is important.

Thus, we need to detect patterns in graphs, and then generate synthetic graphs matching such patterns automatically. This is exactly the focus of the first of four parts of this book. In Part I we discuss graph patterns that appear often in real-world graphs, with more emphasis on static graphs, which are relatively easier to obtain and study. We also report patterns for graphs that evolve over time, as well as patterns for weighted graphs.

In Part II we describe some graph generators which try to match one or more of the above patterns. Typically, we provide the motivation for the generator, the main ideas behind it, and the list

---

[1] https://newsroom.fb.com/content/default.aspx?NewsAreaId=22

Symbol	Description
$\mathcal{G}$	A graph with $(\mathcal{V}, \mathcal{E})$ set of nodes and edges
$\mathcal{V}$	Set of nodes for graph $\mathcal{G}$
$\mathcal{E}$	Set of edges for graph $\mathcal{G}$
N	Number of nodes, or $|\mathcal{V}|$
E	Number of edges, or $|\mathcal{E}|$
$e_{i,j}$	Edge between node i and node j
$w_{i,j}$	Weight on edge $e_{i,j}$
w_i	Weight of node i (sum of weights of incident edges)
$\mathbf{A}$	0-1 Adjacency matrix of the unweighted graph
$\mathbf{A}_w$	Real-value adjacency matrix of the weighted graph
$a_{i,j}$	Entry in matrix $\mathbf{A}$
λ_1	Principal eigenvalue of unweighted graph
$\lambda_{1,w}$	Principal eigenvalue of weighted graph
γ	Power-law exponent: $y(x) \propto x^{-\gamma}$

Figure 1.1: Table of symbols used in notation.

of patterns it satisfies. We describe in more detail the so-called "*RMat*" generator and its follow-up, the "Kronecker graphs," because they have several desirable properties, and they are the basis behind the so-called *graph500* generator[2] that is heavily used for supercomputing benchmarks.

What are the main tools to study large graphs? We answer this question in Part III. There, we give the intuition and some applications of the powerful singular value decomposition (SVD) and the closely related eigenvalue analysis, and we present several applications: They form the basis of the famous PageRank algorithm of Google (Section 14.4); and they have some additional surprising uses, like automatic image captioning and estimating the proximity of two nodes (Section 18.1). Finally, we present a recent result in epidemiology and virus propagation (Chapter 17), which states that, for a given virus, an epidemic can never occur if the first eigenvalue of the connectivity graph satisfies a simple condition.

In the final part, Part IV, we briefly survey related work from machine learning, data mining, visualization, as well as sociology.

Figure 1.1 lists the symbols used in this book. Figure 1.2 gives the description of some of the datasets we will refer to [203].

[2]http://www.graph500.org/

Name	Weights	|N|,|E|,time	Description
PostNet	Unweighted	250K, 218K, 80 d.	Blog post citation network
NIPS	Unweighted	2K, 3K, 13 yr.	Paper citation network
Arxiv	Unweighted	30K, 60K, 13 yr.	Paper citation network
Patent	Unweighted	4M, 8M, 17 yr.	Patent citation network
IMDB	Unweighted	757K, 2M, 114 yr.	Bipartite actor-movie network
Netflix	Unweighted	125K, 14M, 72 mo.	Bipartite user-movie ratings
BlogNet	Multi-edges	60K, 125K, 80 d.	Social network of blogs based on citations
Auth-Conf	Multi-edges	17K, 22K, 25 yr.	Bipartite DBLP author-to-conference associations
Key-Conf	Multi-edges	10K, 23K, 25 yr.	Bipartite DBLP keyword-to-conference associations
Auth-Key	Multi-edges	27K, 189K, 25 yr.	Bipartite DBLP author-to-keyword associations
CampOrg	Edge-weights (Amounts)	23K, 877K, 28 yr.	Bipartite U.S. electoral campaign donations from organizations to candidates (available from FEC)
CampIndiv	Edge-weights (Amounts)	6M, 10M, 22 yr.	Bipartite election donations from individuals to organizations
Twitter	Unweighted	62.5M, 2.78B, 2yr	who follows whom in 2009-11
YahooWeb	Unweighted	1.4B, 6.6B, n/a	WWW pages in 2002 snapshot
Epinions	Unweighted	75K, 508K, n/a	who trusts whom on epinions.com
ClickStream	Unweighted	75K, 508K, n/a	users clicking on web sites
Oregon	Unweighted	4K, 15K	connections across autonomous systems (AS)

Figure 1.2: The datasets referred to in this work.

Dataset Description   We will use the datasets of Figure 1.2 on several occasions in this book. The second column indicates whether they are weighted or not, and whether we allow multiple edges (*multi-edges*) between a pair of nodes, as in the case of person "A" calling person "B" several times during the time-interval of observation. Some of the graphs are *bipartite*, like the movie-actor dataset (who acted in what movie). We use the term *unipartite* for the rest, like, e.g., the *Patents* datasets with patents citing older patents. The third column in Figure 1.2 gives the number of nodes and edges (N and E, respectively), as well as the time interval of observation, for time-evolving graphs.

The datasets are gathered from publicly available data. *NIPS* [3], *Arxiv*, and *Patent* [184] are academic paper or patent citation graphs with no weighting scheme. *IMDB* indicates movie-actor information, where an edge occurs if an actor participates in a movie [35]. *Netflix* is the dataset from the Netflix Prize competition[4], with user-movie links (we ignored the ratings); we also noticed that

[3] www.cs.toronto.edu/~roweis/data.html
[4] www.netflixprize.com

it only contained users with 100 or more ratings. *BlogNet* and *PostNet* are two representations of the same data, hyperlinks between blog posts [188]. In *PostNet*, nodes represent individual posts, while in *BlogNet* each node represents a blog. Essentially, *PostNet* is a paper citation network while *BlogNet* is an author citation network (which contains multi-edges).

Auth-Conf, *Key-Conf*, and *Auth-Key* are all from DBLP[5], with the obvious meanings. *CampOrg* and *CampIndiv* are bipartite graphs from the U.S. Federal Election Commission, recording donation amounts from organizations to political candidates and individuals to organizations[6].

After all these preliminaries, we are ready to tackle the question: *Are real graphs random?*

---

[5]dblp.uni-trier.de/xml/
[6]www.cs.cmu.edu/~mmcgloho/fec/data/fec_data.html

PART I

Patterns and Laws

CHAPTER 2

Patterns in Static Graphs

Are real graphs random? The answer is a resounding "no." In addition to the famous "six degrees of separation" (see Section 2.3 on page 12), there is a long list of rules that real graphs often obey.

In this chapter, we present the "laws" that apply to static snapshots of real graphs, and we ignore the weights on the edges. These patterns include the rules in degree distributions, diameter (the number of hops in which pairs of nodes can reach each other), local number of triangles, eigenvalues, and communities. Next, we describe the related patterns in more detail.

2.1   S-1: HEAVY-TAILED DEGREE DISTRIBUTION

Suppose in a social (or computer) network of, say, 1 million nodes, each node has on the average 50 direct contacts ('friends'). Let's consider the following two questions:

- Q1: If you pick a node at random, what is your best guess for the number of friends it has?
- Q2: Would you be surprised if a node has degree 10,000?

For the first question, most people would respond: "about 50, give or take a little."

For the second question, most people would say 'yes:' such a huge number of friends is suspicious.

Both answers are wrong. For the first question, the correct answer is closer to '1' – most people/nodes are barely connected (no easy access to PC and Internet, too expensive, too confusing, too little benefit for them to connect). For the second question, we should be surprised if we *don't* find a node with 10,000 friends, given that we have 1 million nodes in the network.

What is happening is that, with few exceptions, the overwhelming majority of networks have a very skewed degree distribution, along the lines of the Zipf distribution [289] of vocabulary words in text: some words, like "the" and "a," appear very often, in the same way some nodes in a network have a huge degree (because they are actually a job-placement agency in LinkedIn, or they want to win a popularity contest on Facebook, or they simply enjoy having a lot of friends/contacts).

Conversely, the vast majority of vocabulary words appear once or twice: as a point of reference, the Oxford English Dictionary has over 100,000 words, the Unix(TM) spell-check dictionary has about 30,000[1], while anecdotal observations state that adult native speakers use no more than 2,000 vocabulary words in their every day life. That is, the distribution of word usage is *very* skewed. Similarly, the degree distribution of the nodes in the network is also skewed, with the majority of nodes being barely connected.

---

[1](/usr/dict/words)

Why is our intuition so wrong? There are several opinions (see, e.g., [258]), but it seems that we subconsciously assume that the degree distribution is Gaussian (the famous "bell curve" – see Figure 2.1 (a)). The Gaussian distribution has exactly the properties we expect: most of the nodes should have a degree close to the average, and large deviations from the average are improbable. This intuition seems to hold for random variables like body height, or body weight of patients, but it fails miserably for income, and, as we shall see next, for several graph-related quantities.

One of the many such quantities is the degree of a node. The degree distribution of many real graphs obey a power law of the form $f(d) \propto d^{-\alpha}$, with the exponent $\alpha > 0$, and $f(d)$ being the fraction of nodes with degree d. Such power-law relations as well as many more have been reported in [72, 114, 169, 213]. Intuitively, power-law-like distributions for degrees state that there exist many low degree nodes, whereas only a few high degree nodes exist in real graphs.

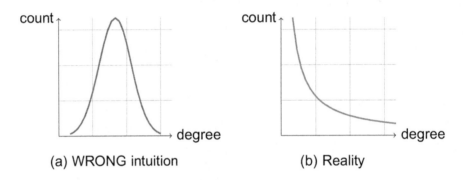

Figure 2.1: Incorrect intuition about degree distributions.

Definition 2.1 Power Law   Two variables x and y are related by a power law when:

$$y(x) = Ax^{-\gamma} \tag{2.1}$$

where A and γ are positive constants.

The constant γ is often called the *power-law exponent*.

Definition 2.2 Power-Law Distribution   A random variable is distributed according to a power law when the probability density function (pdf) is given by:

$$p(x) = Ax^{-\gamma}, \quad \gamma > 1, x \geq x_{min} \tag{2.2}$$

The extra $\gamma > 1$ requirement ensures that $p(x)$ can be normalized. Power laws with $\gamma < 1$ rarely occur in nature, if ever [215].

Skewed distributions, such as power laws, occur very often. In the Internet graph, the degree distribution follows such a power law [115]; that is, the count c_k of nodes with degree k, versus the degree k, is a line on a log-log scale. The World Wide Web graph also obeys power laws [169]: the in-degree and out-degree distributions both follow power laws, as well as the number of the so-called "bipartite cores" ($\approx$ communities, which we will see later) and the distribution of PageRank values [59, 227]. Redner [237] shows that the citation graph of scientific literature follows a power law with exponent 3. Figures 2.2 show examples of such power laws.

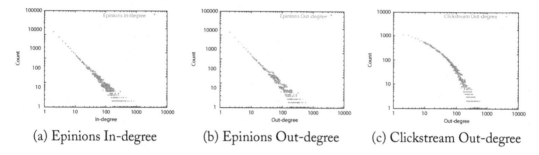

| (a) Epinions In-degree | (b) Epinions Out-degree | (c) Clickstream Out-degree |

Figure 2.2: *Power laws:* Plots (a) and (b) show the in-degree and out-degree distributions on a log-log scale for the *Epinions* graph (an online social network of $75K$ people and $508K$ edges [95]). Plot (c) shows the out-degree distribution of a *ClickStream* graph (a bipartite graph of users and the websites they surf [211]). They all have a skewed distribution.

The significance of a power-law distribution $p(x)$ is that it decays much slower ('heavy tail') as $x \to \infty$, as against the faster decay for the Gaussian and other exponential distributions. Thus, a power-law degree distribution would be much more likely to have nodes with a very high degree (much larger than the mean) than the Gaussian distribution. This observation is at the heart of popular science books, like the book on 'black swans' [258]; and also behind the "mice and elephant" job distribution in queueing theory, the '80-20' rule, the Pareto distribution of income, and many more.

Graphs exhibiting such skewed degree distributions are called *scale-free* graphs, because the form of $y(x)$ in Equation 2.1 remains unchanged to within a multiplicative factor when the variable x is multiplied by a scaling factor (in other words, $y(ax) = by(x)$). Thus, there is no special "characteristic scale" for the variables; the functional form of the relationship remains the same for all scales.

2.2 S-2: EIGENVALUE POWER LAW (EPL)

Real graphs obey several other, surprising regularities. Siganos et al. [247] examined the spectrum of the adjacency matrix of the AS Internet topology and reported that the 20–100 largest eigenvalues [2]

---

[2]Quick reminder: for a square matrix $\mathbf{A}$, the scalar-vector pair $\lambda, \vec{u}$ are called eigenvalue-eigenvector pair, if $\mathbf{A}\vec{u} = \lambda\vec{u}$.

of the Internet graph are power-law distributed. The spectrum, or *scree plot* of a matrix, is the set of its eigenvalues sorted in decreasing magnitude. We give the detailed definition and intuition of eigenvalues later (Section 14.1), along with their usefulness in the study of graphs.

Mihail and Papadimitriou [206] later provided a plausible explanation for the "Eigenvalue Power Law," showing that, under mild assumptions, it is a consequence of the "Degree Power Law."

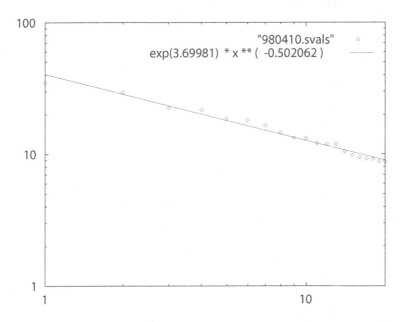

Figure 2.3: Scree plot obeys a power law: Eigenvalue λ_i versus rank i for the adjacency matrix of autonomous systems (April 10, '98).

2.3   S-3 SMALL DIAMETER

One of the most striking patterns that real-world graphs have is a small diameter, which is also known as the "small-world phenomenon" or the "six degrees of separation."

For a given static graph, its diameter is defined as the maximum *distance* between any two nodes, where distance is the minimum number of hops (i.e., edges that must be traversed) on the path from one node to another, usually ignoring directionality. Intuitively, the diameter represents how much of a "small world" the graph is – how quickly one can get from one "end" of the graph to another.

Many real graphs were found to exhibit surprisingly small diameters – for example, 19 for the Web [18], and the well-known "six-degrees of separation" in social networks [36]. It has also been observed that the diameter spikes at the "gelling point" [201].

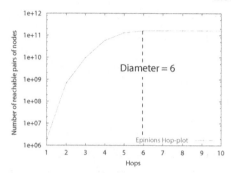

Figure 2.4: *Hop-plot and effective diameter:* This is the hop-plot of the *Epinions* graph [73, 95]. We see that the number of reachable pairs of nodes flattens out at around six hops; thus the effective diameter of this graph is 6.

Since the diameter is defined as the *maximum*-length shortest path between all possible pairs, it can easily be highjacked by long chains. A more robust, and oft-preferred, metric is the *effective diameter*, which is the 90-percentile of the pairwise distances among all reachable pairs of nodes. In other words, the *effective diameter* is the minimum number of hops in which some fraction (usually 90%) of all connected node pairs can be reached [248].

Computing all-pairs-shortest-path lengths is practically intractable for very large graphs. The exact algorithm is prohibitively expensive (at least $O(N^2)$); while one can use sampling to estimate it, alternative methods would include the so-called *approximate neighborhood function* ANF [220].

Informal description: Travers and Milgram [268] conducted a famous experiment where participants were asked to reach a randomly assigned target individual by sending a chain letter. They found that for all the chains that completed, the average length of such chains was six, which is a very small number considering the large population the participants and targets were chosen from. This leads us to believe in the concept of "six degrees of separation:" the diameter of a graph is an attempt to capture exactly this.

Detailed description: Several (often related) terms have been used to describe the idea of the "diameter" of a graph:

- *Expansion and the "hop-plot"*: Tangmunarunkit et al. [259] use a well-known metric from theoretical computer science called "expansion," which measures the rate of increase of neighborhood with increasing h. This has been called the "hop-plot" elsewhere [115].

Definition 2.3 Hop-plot   Starting from a node u in the graph, we find the number of nodes $N_h(u)$ in a neighborhood of h hops. We repeat this starting from each node in the graph, and sum the results to find the total neighborhood size N_h for h hops ($N_h = \sum_u N_h(u)$). The hop-plot is just the plot of N_h versus h.

- *Effective diameter or eccentricity*: The hop-plot helps calculate the *effective diameter* (also called the *eccentricity*) of the graph.

Definition 2.4 Effective diameter   This is the minimum number of hops in which some fraction (typically 90%) of all connected pairs of nodes can reach each other [261]. Figure 2.4 shows the hop-plot and effective diameter of an example graph.

- *Characteristic path length*: For each node in the graph, consider the shortest paths from it to every other node in the graph. Take the average length of all these paths. Now, consider the average path lengths for *all* possible starting nodes, and take their median. This is the characteristic path length [62].

- *Average diameter*: This is calculated in the same way as the characteristic path length, except that we take the mean of the average shortest path lengths over all nodes, instead of the median.

While the use of "expansion" as a metric is somewhat vague (Tangmunarunkit et al. [259] use it only to differentiate between exponential and sub-exponential growth), most of the other metrics are quite similar. The advantage of eccentricity is that its definition works, as is, even for disconnected graphs, whereas we must consider only the largest component for the characteristic and average diameters. Characteristic path length and eccentricity are less vulnerable to outliers than average diameter, but average diameter might be the better if we want worst case analysis.

Computation issues: One major problem with finding the diameter is the computational cost: all the definitions essentially require computing the "neighborhood size" of each node in the graph. One approach is to use repeated matrix multiplications on the adjacency matrix of the graph; however, this takes asymptotically $O(N^{2.37})$ time [282]. Another technique is to do breadth-first searches from each node of the graph. This takes $O(N + E)$ space but requires $O(NE)$ time. Another issue with breadth-first search is that edges are not accessed sequentially, which can lead to terrible performance on disk-resident graphs. Palmer et al. [220] find that randomized breadth-first search algorithms are also ill-suited for large graphs, and they provide a randomized algorithm for finding the hop-plot which takes $O((N + E)d)$ time and $O(N)$ space (apart from the storage for the graph itself), where N is the number of nodes, E the number of edges, and d the diameter of the graph (typically very small). Their algorithm offers provable bounds on the quality of the approximated result, and requires only sequential scans over the data. They find the technique to be far faster than exact computation, and it provides much better estimates than other schemes like sampling.

Examples in the real world: The diameters of several naturally occurring graphs have been calculated, and in almost all cases they are very small compared to the graph size. Faloutsos et al. [115] find an effective diameter of around 4 for the Internet AS level graph and around 12 for the Router level graph. Govindan and Tangmunarunkit [132] find a 97%-effective diameter of around 15 for the Internet Router graph. Broder et al. [61] find that the average path length in the WWW (when a path exists at all) is about 16 if we consider the directions of links, and around 7 if all edges are considered to be undirected. Backstrom et al. [27] find a radius of only 4 for the Facebook social network. Albert et al. [19] find the average diameter of the webpages in the nd.edu domain to be 11.2. Watts and Strogatz [279] find the average diameters of the power grid and the network of actors to be 18.7 and 3.65 respectively. Kang et al. [154] found the diameter of *Yahoo Web* (a billion-node web crawl – see Figure 1.2 on page 4) to be about 7; Leskovec and Horvitz [183] found the diameter of the instant-messenger (IM) network in Microsoft to be also about 7. Many other such examples can be found in the literature; Tables 1 and 2 of [17] and Table 3.1 of [214] list some such work.

2.4   S-4, S-5: TRIANGLE POWER LAWS (TPL, DTPL)

Real social networks tend to have many more triangles than random, because friends of friends often become friends themselves. Let Δ be the number of triangles in a graph, and Δ_i the number of triangles that node i participates in. Tsourakakis [270] discovered two power laws with respect to triangles. The first, *triangle participation law* or TPL, says that the distribution of Δ_i follows a power law with exponent σ. The TPL intuitively states that while many nodes have only a few triangles in their neighborhoods, a few nodes participate in a huge number of triangles. Figure 2.5(a) illustrates the TPL, for the *Epinions* dataset we have seen before (70K nodes, 500K edges, of who-trusts-whom).

The number of triangles Δ_i for node i is related to the *clustering coefficient* of the node (see definition in Section 16.1), and, eventually, to the clustering coefficient of the whole graph.

The second pattern links the degree d_i of a node, with the number of triangles Δ_i that it participates in. Intuitively, we would expect a positive, possibly linear, correlation: the more friends/contacts node i has, the more triangles her friends will form. The correlation is indeed positive, but the exact relationship is *super-linear*, and actually, a power law with positive slope: $\Delta_i \propto d_i^s$, where $s \approx 1.5$. Figure 2.5 gives this plot for the *Epinions* dataset.

Kang et al. [150] studied the *Twitter* graph. Figure 2.6 gives the same DTPL scatter-plot (degree d_i vs Δ_i), for Twitter accounts. We see that celebrities have high degree and reasonably connected followers, and so does the vast majority of typical users, who are omitted for visual clarity. They all seem to follow the super-linear relationship of degree/triangles, with a few glaring exceptions, the cloud at the top-left part of the plot. Those accounts have a relatively small degree $\approx 10^4$, but participate in about 10^8 triangles, which is three orders of magnitude more than that for typical users. Closer inspection shows that these are accounts of adult advertisers, and possibly

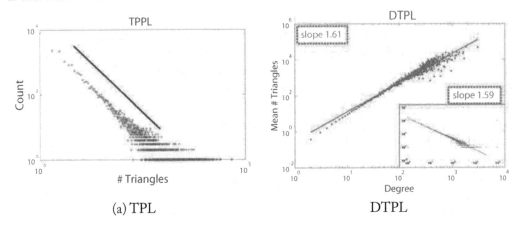

Figure 2.5: Illustration of the triangle laws, on the "*Epinions*" dataset. (a) "triangle participation" law (TPL) – histogram of count of participating triangles; (b) the "degree-triangle" law (DTPL) – scatter plot of degree d_i vs Δ_i, for each node i. All axis logarithmic.

the same person with multiple accounts; these accounts are following each other, to artificially boost their degrees and credibility.

In conclusion, degree-triangle plots can be used to spot potentially dangerous accounts such as those of adult advertisers and spammers.

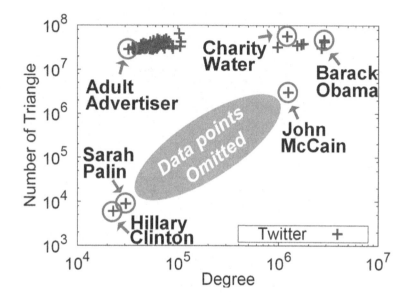

Figure 2.6: The degree d_i vs. participating triangles Δ_i of some "celebrities" (rest: omitted, for clarity) in Twitter accounts. Also shown are accounts of adult sites which have smaller degree, but belong to an abnormally large number of triangles (= many, well-connected followers – probably, "robots").

CHAPTER 3

Patterns in Evolving Graphs

So far we studied static patterns: given a graph, what are the regularities we can observe? Here we study time evolving graphs, such as patents citing each other (and new patents arriving continuously), autonomous system connecting to each other (with new or dropped connections, as time passes), and so on.

3.1   D-1: SHRINKING DIAMETERS

As a graph grows over time, one would intuitively expect that the diameter grows, too. It should be a slow growth, given the "six degrees" of separation. Does it grow as $O(\log N)$? As $O(\log \log N)$? Both guesses sound reasonable.

It turns out that they are both wrong. Surprisingly, graph diameter *shrinks*, even when new nodes are added [191]. Figure 3.1 shows the *Patents* dataset, and specifically its ("effective") diameter over time. (Recall that the effective diameter is the number of hops d such that 90% of all the reachable pairs can reach each other within d or fewer hops). Several other graphs show the same behavior, as reported in [191] and [153].

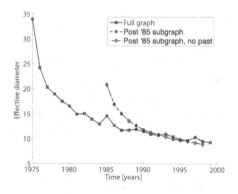

Figure 3.1: Diameter over time – patent citation graph

3.2   D-2: DENSIFICATION POWER LAW (DPL)

Time-evolving graphs present additional surprises. Suppose that we had $N(t)$ nodes and $E(t)$ edges at time t, and suppose that the number of nodes doubled $N(t+1) = 2 * N(t)$ – what is our best

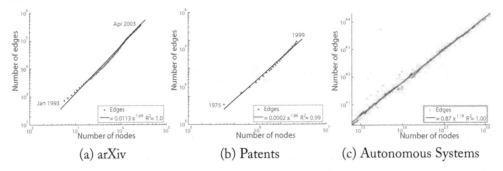

(a) arXiv   (b) Patents   (c) Autonomous Systems

Figure 3.2: *The Densification Power Law:* The number of edges $E(t)$ is plotted against the number of nodes $N(t)$ on log-log scales for (a) the *Arxiv* citation graph, (b) the *Patent* citation graph, and (c) *Oregon*, the Internet Autonomous Systems graph. All of these grow over time, and the growth follows a power law in all three cases [191].

guess for the number of edges $E(t+1)$? Most people will say 'double the nodes, double the edges.' But this is also wrong: the number of edges grows super-linearly to the number of nodes, following a power law, with a positive exponent. Figure 3.2 illustrates the pattern for three different datasets (*Arxiv*, *Patent*, *Oregon*– see Figure 1.2 for their description)

Mathematically, the equation is

$$E(t) \propto N(t)^\beta$$

for all time ticks, where β is the densification exponent, and $E(t)$ and $N(t)$ are the number of edges and nodes at time t, respectively.

All the real graphs studied in [191] obeyed the DPL, with exponents between 1.03 and 1.7. When the power-law exponent $\beta > 1$ then we have a super-linear relationship between the number of nodes and the number of edges in real graphs. That is, when the number of nodes N in a graph doubles, the number of edges E more than doubles – hence the densification. This may also explain why the diameter shrinks: as time goes by, the average degree grows, because there are many more new edges than new nodes.

Before we move to the next observation, one may ask: *Why does the average degree grow?* Do people writing patents cite more patents than ten years ago? Do people writing physics papers cite more papers than earlier? Most of us write papers citing the usual number of earlier papers (10-30) – how come the average degree grows with time?

We conjecture that the answer is subtle, and is based on the power-law degree distribution ($S-1$ pattern): the more we wait, the higher the chances that there will be a super-patent with a huge count of citations, or a survey paper citing 200 papers, or a textbook citing a thousand papers. Thus, the 'mode,' the typical count of citations per paper/patent, remains the same, but the average is hijacked by the (more than one) high-degree newcomers.

This is one more illustration of how counter-intuitive power laws are. If the degree distribution of patents/papers was Gaussian or Poisson, then its variance would be small, the average degree would stabilize toward the distribution mean, and the diameter would grow slowly ($O(\log N)$ or so [80, 195]), but it would not shrink.

3.3   D-3: DIAMETER-PLOT AND GELLING POINT

Studying the effective diameter of the graphs, McGlohon et al. noticed that there is often a point in time when the diameter spikes [200, 201]. Before that point, the graph typically consists of a collection of small, disconnected components. This "*gelling point*" seems to also be the time where the *Giant Connected Component* (GCC) forms and "takes off," in the sense that the vast majority of nodes belong to it, and, as new nodes appear, they mainly tend to join the GCC, making it even larger. We shall refer to the rest of the connected components as "NLCC" (*non-largest connected components*).

Observation 3.1   Gelling point   Real time-evolving graphs exhibit a gelling point, at which the diameter spikes and (several) disconnected components gel into a giant component.

After the gelling point, the graph obeys the expected rules, such as the densification power law; its diameter decreases or stabilizes; and, as we said, the giant connected component keeps growing, absorbing the vast majority of the newcomer nodes.

We show full results for *PostNet* in Fig. 3.3, including the diameter plot (Fig. 3.3(a)), sizes of the NLCCs (Fig. 3.3(b)), densification plot (Fig. 3.3(c)), and the sizes of the three largest connected components in log-linear scale, to observe how the GCC dominates the others (Fig. 3.3(d)). Results from other networks are similar, and are shown in condensed form for brevity (Fig. 3.4). The left column shows the diameter plots, and the right column shows the NLCCs, which also present some surprising regularities, that we describe next.

3.4   D-4: OSCILLATING NLCCS SIZES

After the gelling point, the giant connected component (GCC) keeps on growing. What happens to the 2nd, 3rd, and other connected components (the NLCCs)?

- Do they grow with a smaller rate, following the GCC?
- Do they shrink and eventually get absorbed into the GCC?
- Or do they stabilize in size?

It turns out that they do a little bit of all three of the above: in fact, they *oscillate* in size.

Further investigation shows that the oscillation may be explained as follows: new comer nodes typically link to the GCC; very few of the newcomers link to the 2nd (or 3rd) CC, helping them to grow slowly; in very rare cases, a newcomer links both to an NLCC, as well as the GCC, thus leading to the absorption of the NLCC into the GCC. It is exactly at these times that we have a drop in the size of the 2nd CC. Note that edges are not removed; thus, what is reported as the size of the 2nd CC is actually the size of yesterday's 3rd CC, causing the apparent "oscillation."

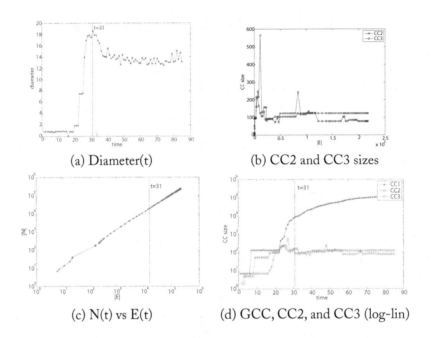

(a) Diameter(t)

(b) CC2 and CC3 sizes

(c) N(t) vs E(t)

(d) GCC, CC2, and CC3 (log-lin)

Figure 3.3: Properties of *PostNet* network. Notice that we experience an early gelling point at (a) (diameter versus time), stabilization/oscillation of the NLCC sizes in (b) (size of 2nd and 3rd CC, versus time). The vertical line marks the gelling point. Part (c) gives $N(t)$ vs $E(t)$ in log-log scales – the good linear fit agrees with the Densification Power Law. Part (d): component size (in log), vs time – the GCC is included, and it clearly dominates the rest, after the gelling point.

A rather surprising observation is that the largest size of these components seems to be a constant over time.

The second column of Fig. 3.4 show the NLCC sizes versus time. Notice that, after the "gelling" point (marked with a vertical line), they all oscillate about a constant value (different for each network). The actor-movie dataset *IMDB* is especially interesting: the gelling point is around 1914, which is reasonable, given that (silent) movies started around 1890: Initially, there were few movies and few actors who didn't have the chance to play in the same movie. After a few years, most actors played in at least one movie with some other, well-connected actor, and hence the giant

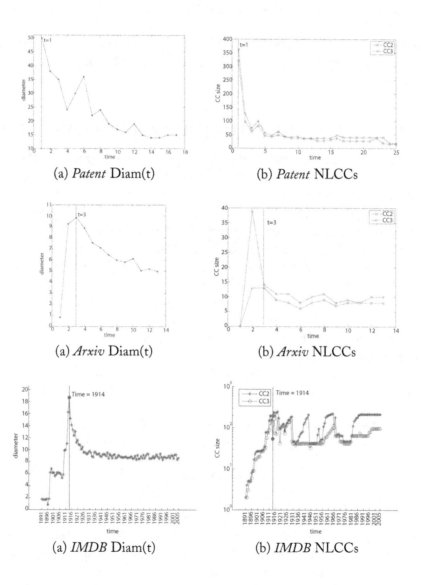

(a) *Patent* Diam(t)   (b) *Patent* NLCCs

(a) *Arxiv* Diam(t)   (b) *Arxiv* NLCCs

(a) *IMDB* Diam(t)   (b) *IMDB* NLCCs

Figure 3.4: Properties of other networks (*Arxiv*, *Patent*, and *IMDB*). Diameter plot (left column), and NLCCs over time (right); vertical line marks the gelling point. All datasets exhibit an early gelling point, and oscillation of the NLCCs.

connected component (GCC) was formed. After that, the size of the 2nd NLCC seems to oscillate between 40 and 300, with growth periods followed by sudden drops (= absorptions into GCC).

Observation 3.2 Oscillating NLCCs After the gelling point, the secondary and tertiary connected components remain of approximately constant size, with relatively small oscillations (relative to the size of the GCC).

3.5 D-5: LPL: PRINCIPAL EIGENVALUE OVER TIME

A very important measure of connectivity of a graph is its first eigenvalue[1]. We mentioned eigenvalues earlier under the Eigenvalue Power Law (Section 2.2), and we reserve a fuller discussion of eigenvalues and singular values for later (Section 14.1 and Section 14.2). Here, we focus on the principal (=maximum magnitude) eigenvalue λ_1. This is an especially important measure of the connectivity of the graph, effectively determining the so-called *epidemic threshold* for a virus, as we discuss in Chapter 17. For the time being, the intuition behind the principal eigenvalue λ_1 is that it roughly corresponds to the average degree. In fact, for the so-called *homogeneous* graphs (all nodes have the same degree), it *is* the average degree.

McGlohon et al. [200, 201], studied the principal eigenvalue λ_1 of the 0-1 adjacency matrix **A** of several datasets over time. As more nodes (and edges) are added to a graph, one would expect the connectivity to become better, and thus λ_1 should grow. Should it grow linearly with the number of edges E? Super-linearly? As $O(\log E)$?

They notice that the principal eigenvalue seems to follow a power law with increasing number of edges E. The power-law fit is better after the *gelling point* (Observation 3.1).

Observation 3.3 λ_1 Power Law (LPL) In real graphs, the principal eigenvalue $\lambda_1(t)$ and the number of edges $E(t)$ over time follow a power law with exponent less than 0.5. That is,

$$\lambda_1(t) \propto E(t)^{\alpha}, \ \alpha \leq 0.5$$

Fig. 3.5 shows the corresponding plots for some networks, and the power-law exponents. Note that we fit the given lines *after* the gelling point, which is indicated by a vertical line for each dataset. Also note that the estimated slopes are less than 0.5, which is in agreement with graph theory: The most connected unweighted graph with N nodes would have $E = N^2$ edges, and eigenvalue $\lambda_1 = N$; thus $\log \lambda_1 / \log E = \log N/(2 \log N) = 2$, which means that 0.5 is an extreme value – see [11] for details.

---

[1]Reminder: for a square matrix **A**, the scalar-vector pair $\lambda, \vec{u}$ are called eigenvalue-eigenvector pair, if $\mathbf{A}\vec{u} = \lambda\vec{u}$. See Section 14.1.

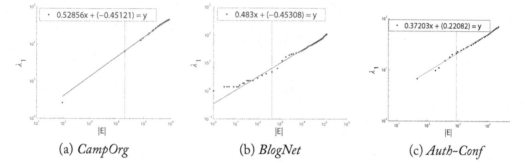

(a) *CampOrg*   (b) *BlogNet*   (c) *Auth–Conf*

Figure 3.5: Illustration of the LPL. 1st eigenvalue $\lambda_1(t)$ of the *0–1* adjacency matrix $\mathbf{A}$ versus number of edges $E(t)$ over time. The vertical lines indicate the gelling point.

CHAPTER 4

Patterns in Weighted Graphs

Here we try to find patterns that weighted graphs obey. The dataset consist of quadruples, such as (IP-source, IP-destination, timestamp, number-of-packets), where timestamp is in increments of, say, 30 minutes. Thus, we have multi-edges, as well as total weight for each (source, destination) pair. Let $W(t)$ be the total weight up to time t (i.e., the grand total of all exchanged packets across all pairs), $E(t)$ the number of distinct edges up to time t, and $n(t)$ the number of multi-edges up to time t.

Following McGlohon et al. [200, 202], we present three "laws" that several datasets seem to follow. The first is the "snapshot power law" (SPL), also known as "fortification," correlating the in-degree with the in-weight, and the out-degree with the out-weight, for all the nodes of a graph at a given time-stamp. The other two laws are for time-evolving graphs: the first, Weight Power Law (WPL), relates the total graph weight to the total count of edges; and the next, Weighted Eigenvalue Power Law (WLPL), gives a power law of the first eigenvalue over time.

4.1 W-1: SNAPSHOT POWER LAWS (SPL)—"FORTIFICATION"

If node i has out-degree out_i, what can we say about its out-weight $outw_i$? It turns out that there is a "fortification effect" here, resulting in more power laws, both for out-degrees/out-weights as well as for in-degrees/in-weights.

Specifically, at a given point in time, we plot the scatterplot of the in/out weight versus the in/out degree, for all the nodes in the graph, at a given time snapshot. An example of such a plot is in Fig. 4.1 (a) and (b). Here, every point represents a node, and the x and y coordinates are its degree and total weight, respectively. To achieve a good fit, we bucketize the x axis with logarithmic binning [213], and, for each bin, we compute the median y.

We observed that the median values of weights versus mid-points of the intervals follow a power law for all datasets studied. Formally, the "Snapshot Power Law" is:

Observation 4.1 Snapshot Power Law (SPL) Consider the i-th node of a weighted graph, at time t, and let $out_i, outw_i$ be its out-degree and out-weight. Then

$$outw_i \propto out_i^{ow}$$

where ow is the *out-weight-exponent* of the SPL. Similarly, for the in-degree, with in-weight-exponent iw.

We don't show it here, but it is interesting to note that [200, 202] report that the SPL exponents of a graph remain almost constant over time.

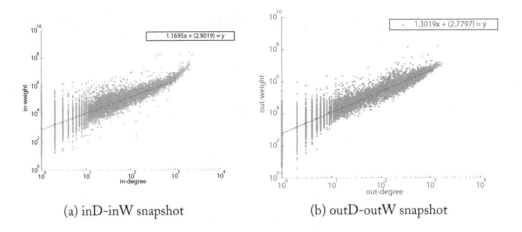

(a) inD-inW snapshot   (b) outD-outW snapshot

Figure 4.1: Snapshot power law (SPL): Weight properties of *CampOrg* donations: (a) and (b) have slopes > 1 ("fortification effect"), that is, the more campaigns an organization supports, the more money it donates (superlinearly), and similarly, the more donations a candidate gets, the higher the amount received.

4.2   DW-1: WEIGHT POWER LAW (WPL)

Recall the definitions of $E(t)$ and $W(t)$: $E(t)$ is the total unique edges up to time t (i.e., count of pairs that know each other) and $W(t)$ is the total count of packets (phonecalls, SMS messages) up to time t. Is there a relationship between $W(t)$ and $E(t)$? If every pair always generated k packets, the relationships would be linear: if the count of pairs double, the packet count would double, too. This is reasonable, but it doesn't happen! In reality, the packet count grows faster, following the "WPL" below.

Observation 4.2   Weight Power Law (WPL)   Let $E(t)$, $W(t)$ be the number of edges and total weight of a graph, at time t. Then, they follow a power law

$$W(t) = E(t)^w \quad (t = 1, 2, \ldots)$$

where w is the *weight* exponent.

Fig. 4.2 demonstrates the WPL for several datasets. The plots are all in log-log scales, and straight lines fit well. The weight exponent w ranges from 1.01 to 1.5 for many real graphs [200]. The highest value corresponds to campaign donations: super-active organizations that support many campaigns also tend to spend even more money per campaign than the less active organizations. In fact, power-laws link the number of nodes $N(t)$ and the number of multi-edges $n(t)$ to $E(t)$ as well.

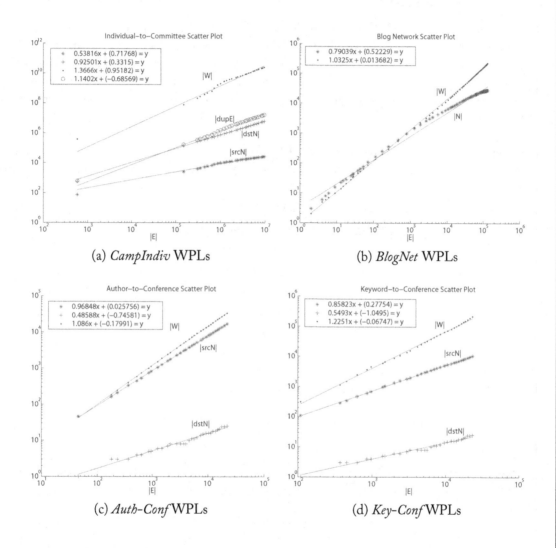

(a) *CampIndiv* WPLs

(b) *BlogNet* WPLs

(c) *Auth-Conf* WPLs

(d) *Key-Conf* WPLs

Figure 4.2: (WPL): Properties of weighted networks. We show total weight $W(t)$ (and count of multi-edges $E_d(t)$ and nodes $N(t)$), versus total count of (unique) edges $E(t)$. Every dot corresponds to a timestamp. The slopes for weight $W(t)$ and multi-edges $E_d(t)$ are above 1, indicating super-linear growth.

4.3   DW-2: LWPL: WEIGHTED PRINCIPAL EIGENVALUE OVER TIME

Given that unweighted (0-1) graphs follow the λ_1 Power Law (LPL, pattern D-5), one may ask if there is a corresponding law for weighted graphs. The answer is 'yes:' let $\lambda_{1,w}$ be the largest eigenvalue of the *weighted* adjacency matrix $\mathbf{A}_w$, where the entries $w_{i,j}$ of $\mathbf{A}_w$ represent the actual edge weight between node i and j (e.g., count of phone-calls from i to j). Notice that $\lambda_{1,w}$ increases with the number of edges, following a power law with a higher exponent than that of its λ_1 Power Law (see Fig. 4.3).

Observation 4.3 $\lambda_{1,w}$ Power Law (LWPL)   Weighted real graphs exhibit a power law for the largest eigenvalue of the weighted adjacency matrix $\lambda_{1,w}(t)$ and the number of edges $E(t)$ over time. That is,

$$\lambda_{1,w}(t) \propto E(t)^\beta$$

In the experiments in [200, 202], the exponent β ranged from 0.5 to 1.6.

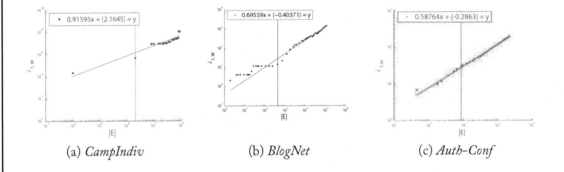

(a) *CampIndiv*   (b) *BlogNet*   (c) *Auth-Conf*

Figure 4.3:   Illustration of the LWPL. 1st eigenvalue $\lambda_{1,w}(t)$ of the *weighted* adjacency matrix $\mathbf{A}_w$ versus number of edges $E(t)$ over time. The vertical lines indicate the gelling point.

CHAPTER 5

Discussion—The Structure of Specific Graphs

While most graphs found naturally share many features (such as the small-world phenomenon), there are some specifics associated with each. These might reflect properties or constraints of the domain to which the graph belongs. We will discuss some well-known graphs and their specific features below.

5.1 THE INTERNET

The networking community has studied the structure of the Internet for a long time. In general, it can be viewed as a collection of interconnected routing domains; each domain is a group of nodes (such as routers, switches, etc.) under a single technical administration [65]. These domains can be considered as either a *stub* domain (which only carries traffic originating or terminating in one of its members) or a *transit* domain (which can carry any traffic). Example stubs include campus networks, or small interconnections of Local Area Networks (LANs). An example transit domain would be a set of backbone nodes over a large area, such as a wide-area network (WAN).

The basic idea is that stubs connect nodes locally, while transit domains interconnect the *stubs*, thus allowing the flow of traffic between nodes from different stubs (usually distant nodes). This imposes a *hierarchy* in the Internet structure, with transit domains at the top, each connecting several stub domains, each of which connects several LANs.

Apart from hierarchy, another feature of the Internet topology is its apparent *Jellyfish* structure at the AS level (Figure 5.1), found by Tauro et al. [261]. This consists of:

- *A core*, consisting of the highest-degree node and the clique it belongs to; this usually has 8–13 nodes.
- *Layers around the core*, organized as concentric circles around the core; layers further from the core have lower importance.
- *Hanging nodes*, representing one-degree nodes linked to nodes in the core or the outer layers. The authors find such nodes to be a large percentage (about 40–45%) of the graph.

5.2 THE WORLD WIDE WEB (WWW)

Broder et al. [61] find that the Web graph is described well by a "bowtie" structure (Figure 5.2(a)). They find that the Web can be broken into four approximately equal-sized pieces. The core of the

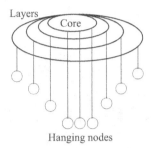

Figure 5.1: *The Internet as a "Jellyfish:"* The Internet AS-level graph can be thought of as a core, surrounded by concentric layers around the core. There are many one-degree nodes that hang off the core and each of the layers.

bowtie is the *Strongly Connected Component* (SCC) of the graph: each node in the SCC has a directed path to any other node in the SCC. Then, there is the IN component: each node in the IN component has a directed path to all the nodes in the SCC. Similarly, there is an OUT component, where each node can be reached by directed paths from the SCC. Apart from these, there are webpages which can reach some pages in OUT and can be reached from pages in IN without going through the SCC; these are the TENDRILS. Occasionally, a tendril can connect nodes in IN and OUT; the tendril is called a TUBE in this case. The remainder of the webpages fall in *disconnected components*. A similar study focused on only the Chilean part of the Web graph found that the disconnected component is actually very large (nearly 50% of the graph size) [31].

Dill et al. [93] extend this view of the Web by considering subgraphs of the WWW at different scales (Figure 5.2(b)). These subgraphs are groups of webpages sharing some common trait, such as content or geographical location. They have several remarkable findings:

1. *Recursive bowtie structure*: Each of these subgraphs forms a bowtie of its own. Thus, the Web graph can be thought of as a hierarchy of bowties, each representing a specific subgraph.
2. *Ease of navigation*: The SCC components of all these bowties are tightly connected together via the SCC of the whole Web graph. This provides a navigational backbone for the Web: starting from a webpage in one bowtie, we can click to its SCC, then go via the SCC of the entire Web to the destination bowtie.
3. *Resilience*: The union of a random collection of subgraphs of the Web has a large SCC component, meaning that the SCCs of the individual subgraphs have strong connections to other SCCs. Thus, the Web graph is very resilient to node deletions and does not depend on the existence of large taxonomies such as yahoo.com; there are several alternate paths between nodes in the SCC.

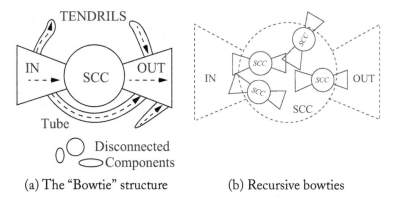

(a) The "Bowtie" structure   (b) Recursive bowties

Figure 5.2: *The "Bowtie" structure of the Web*: Plot (a) shows the four parts: IN, OUT, SCC, and TENDRILS [61]. Plot (b) shows *Recursive Bowties*: subgraphs of the WWW can each be considered a bowtie. All these smaller bowties are connected by the navigational backbone of the main SCC of the Web [93].

CHAPTER 6

Discussion—Power Laws and Deviations

6.1 POWER LAWS—SLOPE ESTIMATION

We saw many power laws in the previous sections. Here we describe how to estimate the slope of a power law, and how to estimate the goodness of fit. We discuss these issues below, using the detection of power laws in degree distributions as an example.

Computing the power-law exponent: This is no simple task: the power law could be only in the tail of the distribution and not over the entire distribution, estimators of the power-law exponent could be biased, some required assumptions may not hold, and so on. Several methods are currently employed, though there is no clear "winner" at present.

1. *Linear regression on the log-log scale:* We could plot the data on a log-log scale, then optionally "bin" them into equal-sized buckets, and finally find the slope of the linear fit. However, there are at least three problems: (i) this can lead to biased estimates [130], (ii) sometimes the power law is only in the *tail* of the distribution, and the point where the tail begins needs to be hand-picked, and (iii) the right end of the distribution is very noisy [215]. However, this is the simplest technique, and seems to be the most popular one.

2. *Linear regression after logarithmic binning:* This is the same as above, but the bin widths increase exponentially as we go toward the tail. In other words, the number of data points in each bin is counted, and then the height of each bin is divided by its width to normalize. Plotting the histogram on a log-log scale would make the bin sizes equal, and the power law can be fitted to the heights of the bins. This reduces the noise in the tail buckets, fixing problem (iii). However, binning leads to loss of information; all that we retain in a bin is its average. In addition, issues (i) and (ii) still exist.

3. *Regression on the cumulative distribution:* We convert the pdf $p(x)$ (that is, the scatter plot) into a *cumulative distribution* $F(x)$:

$$F(x) = P(X \geq x) = \sum_{z=x}^{\infty} p(z) = \sum_{z=x}^{\infty} A z^{-\gamma} \qquad (6.1)$$

The approach avoids the loss of data due to averaging inside a histogram bin. To see how the plot of $F(x)$ versus x will look like, we can bound $F(x)$:

$$\int_x^\infty Az^{-\gamma}dz < F(x) < Ax^{-\gamma} + \int_x^\infty Az^{-\gamma}dz$$

$$\Rightarrow \quad \frac{A}{\gamma - 1}x^{-(\gamma-1)} < F(x) < Ax^{-\gamma} + \frac{A}{\gamma - 1}x^{-(\gamma-1)}$$

$$\Rightarrow \quad F(x) \sim x^{-(\gamma-1)} \qquad (6.2)$$

Thus, the cumulative distribution follows a power law with exponent $(\gamma - 1)$. However, successive points on the cumulative distribution plot are not mutually independent, and this can cause problems in fitting the data.

4. *Maximum-Likelihood Estimator (MLE):* This chooses a value of the power law exponent γ such that the likelihood that the data came from the corresponding power-law distribution is maximized. Goldstein et al. [130] find that it gives good unbiased estimates of γ.

5. *The Hill statistic:* Hill [143] gives an easily computable estimator, that seems to give reliable results [215]. However, it also needs to be told where the tail of the distribution begins.

6. *Fitting only to extreme-value data:* Feuerverger and Hall [116] propose another estimator which is claimed to reduce bias compared to the Hill statistic without significantly increasing variance. Again, the user must provide an estimate of where the tail begins, but the authors claim that their method is robust against different choices for this value.

7. *Non-parametric estimators:* Crovella and Taqqu [89] propose a non-parametric method for estimating the power-law exponent without requiring an estimate of the beginning of the power-law tail. While there are no theoretical results on the variance or bias of this estimator, the authors empirically find that accuracy increases with increasing dataset size, and that it is comparable to the Hill statistic.

Checking for goodness of fit: The correlation coefficient has typically been used as an informal measure of the goodness of fit of the degree distribution to a power law. Recently, there has been some work on developing statistical "hypothesis testing" methods to do this more formally. Beirlant et al. [42] derive a bias-corrected Jackson statistic for measuring goodness of fit of the data to a generalized Pareto distribution. Goldstein et al. [130] propose a Kolmogorov-Smirnov test to determine the fit. Such measures need to be used more often in the empirical studies of graph datasets.

6.2 DEVIATIONS FROM POWER LAWS

We saw several examples of power laws, and there are even more that we didn't cover. Such examples include the Internet AS[1] graph with exponent $2.1 - 2.2$ [115], the in-degree and out-degree dis-

---

[1] Autonomous System, typically consisting of many routers administered by the same entity.

tributions of subsets of the WWW with exponents 2.1 and $2.38 - 2.72$ respectively [37, 61, 179], the in-degree distribution of the African web graph with exponent 1.92 [48], a citation graph with exponent 3 [237], distributions of website sizes and traffic [4], and many others. Newman [215] provides a long list of such works.

One may wonder: is every distribution a power law? If not, are there deviations? The answer is that, yes, there are deviations. In log-log scales, sometimes a parabola fits better, or some more complicated curves fit better. For example, Pennock et al. [231], and others, have observed deviations from a pure power-law distribution in several datasets. Common deviations are exponential cutoffs, the so-called "lognormal" distribution, and the "doubly-Pareto-lognormal" distribution. We briefly cover them all, next.

6.2.1   EXPONENTIAL CUTOFFS

Sometimes the distribution looks like a power law over the lower range of values along the x-axis, but decays very quickly for higher values. Often, this decay is exponential, and this is usually called an exponential cutoff:

$$y(x = k) \propto e^{-k/\kappa} k^{-\gamma} \tag{6.3}$$

where $e^{-k/\kappa}$ is the exponential cutoff term and $k^{-\gamma}$ is the power-law term. Amaral et al. [23] find such behaviors in the electric power-grid graph of Southern California and the network of airports, the vertices being airports and the links being non-stop connections between them. They offer two possible explanations for the existence of such cutoffs. One: high-degree nodes might have taken a long time to acquire all their edges and now might be "aged," and this might lead them to attract fewer new edges (for example, older actors might act in fewer movies). Two: high-degree nodes might end up reaching their "capacity" to handle new edges; this might be the case for airports where airlines prefer a small number of high-degree hubs for economic reasons, but are constrained by limited airport capacity.

6.2.2   LOGNORMALS OR THE "DGX" DISTRIBUTION

Pennock et al. [231] recently found while the whole WWW does exhibit power-law degree distributions, subsets of the WWW (such as university homepages and newspaper homepages) deviate significantly. They observed unimodal distributions on the log-log scale. Similar distributions were studied by Bi et al. [46], who found that a discrete truncated lognormal (called the Discrete Gaussian Exponential or "DGX" by the authors) gives a very good fit. A lognormal is a distribution whose logarithm is a Gaussian; its pdf (probability density function) looks like a parabola in log-log scales. The DGX distribution extends the lognormal to discrete distributions (which is what we get in degree distributions), and can be expressed by the formula:

$$y(x = k) = \frac{A(\mu, \sigma)}{k} \exp\left[-\frac{(\ln k - \mu)^2}{2\sigma^2}\right] \quad k = 1, 2, \dots \tag{6.4}$$

where μ and σ are parameters and $A(\mu, \sigma)$ is a constant (used for normalization if $y(x)$ is a probability distribution). The DGX distribution has been used to fit the degree distribution of a bipartite "clickstream" graph linking websites and users (Figure 2.2(c)), telecommunications, and other data.

6.2.3 DOUBLY-PARETO LOGNORMAL (*DPLN*)

Another deviation is well modeled by the so-called Doubly Pareto Lognormal (*dPln*). Mitzenmacher [210] obtained good fits for file size distributions using *dPln*. Seshadri et al. [245] studied the distribution of phone calls per customer, and also found it to be a good fit. We will describe the results of Seshadri et al. below.

Informally, a random variable that follows the dPln distribution looks like the plots of Figure 6.1: in log-log scales, the distribution is approximated by two lines that meet in the middle of the plot. More specifically, Figure 6.1 shows the empirical pdf (that is, the density histogram) for a switch in a telephone company, over a time period of several months. Plot (a) gives the distribution of the number of distinct partners ("callees") per customer. The overwhelming majority of customers called only one person; until about 80-100 "callees," a power law seems to fit well; but after that, there is a sudden drop, following a power-law with a different slope. This is exactly the behavior of the *dPln*: piece-wise linear, in log-log scales. Similarly, Figure 6.1(b) shows the empirical pdf for the count of phone calls per customer: again, the vast majority of customers make just one phone call, with a piece-wise linear behavior, and the "knee" at around 200 phone calls. Figure 6.1(c) shows the empirical pdf for the count of minutes per customer. The qualitative behavior is the same: piece-wise linear, in log-log scales. Additional plots from the same source ([245]) showed similar behavior for several other switches and several other time intervals. In fact, the dataset in [245] included four switches, over month-long periods; each switch recorded calls made to and from callers who were physically present in a contiguous geographical area.

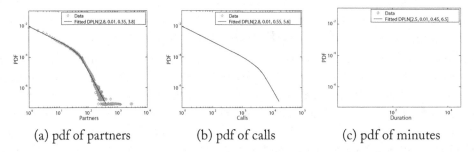

(a) pdf of partners   (b) pdf of calls   (c) pdf of minutes

Figure 6.1: Results of using *dPln* to model. (a) the number of call-partners, (b) the number of calls made, and (c) total duration (in minutes) talked, by users at a telephone-company switch, during a given the time period.

For each customer, the following counts were computed:

- **Partners:** The total number of *unique* callers and callees associated with every user. Note that this is essentially the degree of nodes in the (undirected and unweighted) social graph, which has an edge between two users if either called the other.

- **Calls:** The total number of calls made or received by each user. In graph theoretic terms, this is the weighted degree in the social graph where the weight of an edge between two users is equal to the number of calls that involved them both.

- **Duration:** The total duration of calls for each customer in minutes. This is the weighted degree in the social graph where the weight of the edge between two users is the total duration of the calls between them.

The dPln Distribution   For the intuition and the fitting of the *dPln* distribution, see the work of Reed [238]. Here, we mention only the highlights.

If a random variable X follows the four-parameter *dPln* distribution $dPln(\alpha, \beta, \nu, \tau)$, then the complete distribution is given by:

$$f_X \quad = \frac{\alpha\beta}{\alpha+\beta}\left[e^{(\alpha\nu+\alpha^2\tau^2/2)} x^{-\alpha-1} \Phi\left(\frac{\log x - \nu - \alpha\tau^2}{\tau}\right) + \right.$$
$$\left. x^{\beta-1} e^{(-\beta\tau+\beta^2\tau^2/2)} \Phi^c\left(\frac{\log x - \nu + \beta\tau^2}{\tau}\right)\right], \tag{6.5}$$

where Φ and Φ^c are the CDF and complementary CDF of $N(0, 1)$ (Gaussian with zero mean and unit variance). Intuitively, the pdf looks like a piece-wise linear curve in log-log scales; the parameter α and β are the slopes of the (asymptotically) linear parts, and ν and τ determine the knee of the curve.

One easier way of understanding the double-Pareto nature of X is by observing that X can be written as $X = S_0 \frac{V_1}{V_2}$ where S_0 is lognormally distributed, and V_1 and V_2 are Pareto distributions with parameters α and β. Note that X has a mean that is finite only if $\alpha > 1$ in which case it is given by

$$\frac{\alpha\beta}{(\alpha-1)(\beta+1)} e^{\nu + \frac{\tau^2}{2}}.$$

In conclusion, the distinguishing features of the *dPln* distribution are two linear sub-plots in the log-log scale and a hyperbolic middle section.

CHAPTER 7

Summary of Patterns

In summary, real graphs exhibit surprising, counter-intuitive patterns. In our opinion, the most striking ones are: the "six degrees" or "small world" phenomenon; the skewed degree distribution (also called "scale free," or "power-law tail"); the shrinking diameter; and the higher-than-expected number of triangles.

Here we list the patterns that we presented above. We would like to caution the reader that this list will only keep growing, as more and more researchers focus on real graphs.

- Static graphs (Chapter 2)

  - S-1: Skewed/power-law degree distribution [114][247]
  - S-2: Power-law eigenvalue distribution [114][247]
  - S-3: Small diameter [268]
  - S-4: Triangle power law [269]
  - S-5: Triangle-degree power law [269]

- Dynamic graphs (Chapter 3)

  - D-1: Shrinking diameter [185]
  - D-2: Densification power law [185]
  - D-3: Jelling point [201]
  - D-4: Oscillating size of NLCC (non-largest connected components) [201]
  - D-5: Principal eigenvalue over time [201]

- Weighted Static Graphs (Chapter 4)

  - W-1 superlinear weight-degree relationship ('snapshot power law') [201]

- Weighted Dynamic Graphs (Chapter 4)

  - DW-1: WPL – Weight Power Law. Super-linear relationships, over time between total graph weight $W(t)$, total edge count $E(t)$ etc. [201]
  - DW-2: WLPL – principal eigenvalue of weighted graph follows a power law over time, with respect to edge count [201]

There are several other fascinating observations, like power laws with respect to the number of participating cliques [102]; the "rebel" probability (that is, the probability that a newcomer will *not* connect to the giant-connected component [149]; the bursty nature of edge additions [199]; the power-law drop of blog popularity [189]; the log-logistic distribution of phone call (= edge) durations [90]; patterns with respect to reciprocity [121], [12]; and many more. And, as we said, the list will only continue growing.

With this list of properties, we can tackle the next challenge, namely, how to generate synthetic graphs that mimic the properties of real graphs, as best as we can.

PART II

Graph Generators

All models are wrong, but some are useful

attributed to George E.P. Box

Progress does not involve replacing one
theory that is wrong with one that is right,
rather it involves replacing one theory that is
wrong with one that is more subtly wrong.

attributed to David Hawkins

CHAPTER 8

Graph Generators

Graph generators allow us to create synthetic graphs, which can then be used for, say, simulation studies. But when is such a generated graph "realistic?" This happens when the synthetic graph matches all (or at least several) of the patterns mentioned in the previous chapters. Graph generators can provide insight into graph creation, by telling us which processes can (or cannot) lead to the development of certain patterns.

Graph models and generators can be broadly classified into five categories (Figure 8.1):

1. *Random graph models:* The graphs are generated by a random process. The basic random graph model has attracted a lot of research interest due to its phase transition properties.

2. *Preferential attachment models:* In these models, the "rich" get "richer" as the network grows, leading to power-law effects. Some of today's most popular models belong to this class.

3. *Optimization-based models:* Here, power laws are shown to evolve when risks are minimized using limited resources. Together with the preferential attachment models, they try to provide mechanisms that automatically lead to power laws.

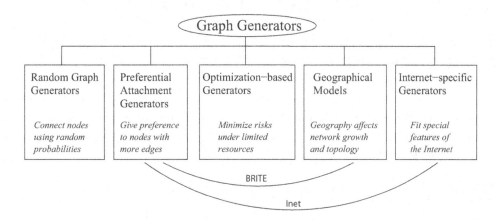

Figure 8.1: *Overview of graph generators:* Current generators can be mostly placed under one of these categories, though there are some hybrids such as *BRITE* and *Inet*.

4. *Geographical models:* These models consider the effects of geography (i.e., the *positions* of the nodes) on the growth and topology of the network. This is especially important for modeling router or power-grid networks, which involve laying wires between points on the globe.

5. *Internet-specific models:* As the Internet is one of the most important graphs in computer science, special-purpose generators have been developed to model its special features. These are often hybrids, using ideas from the other categories and melding them with Internet-specific requirements.

We will discuss graph generators from each of these categories in Sections 8.1–10.3. This is not a complete list, but we believe it includes most of the key ideas from the current literature. Section 10.4 presents work on comparing these graph generators. In Section 11, we discuss the *RMat* generator, which matches many of the patterns mentioned above, and is the basis behind the *graph500* generator[1] used in supercomputing benchmarks. For each generator, we will try to provide the specific problem it aims to solve, followed by a brief description of the generator itself and its properties, and any open questions. Tables 8.1 and 8.2 provide a taxonomy of these.

For the rest of the discussion, we shall use the notation of Figure 1.1 (p. 3), as well as some new symbols, listed on Table 8.3.

8.1   RANDOM GRAPH MODELS

Random graphs are generated by picking nodes under some random probability distribution and then connecting them by edges. We first look at the basic Erdös-Rényi model, which was the first to be studied thoroughly [107], and then we discuss modern variants of the model.

8.1.1   THE ERDÖS-RÉNYI RANDOM GRAPH MODEL

Problem being solved: Graph theory owes much of its origins to the pioneering work of Erdös and Rényi in the 1960s [107, 108]. Their random graph model was the first and the simplest model for generating a graph.

Description and properties: We start with N nodes, and for every pair of nodes, an edge is added between them with probability p (as in Figure 8.2). This defines a *set* of graphs $G_{N,p}$, all of which have the same parameters (N, p).

Degree Distribution: The probability of a vertex having degree k is

$$p_k = \binom{N}{k} p^k (1-p)^{N-k} \approx \frac{z^k e^{-z}}{k!} \quad \text{with } z = p(N-1) \tag{8.1}$$

---

[1]http://www.graph500.org/

Table 8.1: *Taxonomy of graph generators*: This table shows the graph types and degree distributions that different graph generators can create. The graph type can be undirected, directed, bipartite, allowing self-loops, or multi-graph (multiple edges possible between nodes). The degree distributions can be power law (with possible exponential cutoffs, or other deviations such as lognormal/DGX) or exponential decay. If it can generate a power law, the possible range of the exponent γ is provided. Empty cells indicate that the corresponding property does not occur in the corresponding model.

Generator	Undir.	Dir.	Bip.	Self loops	Mult. edges	Geog. info	Power law — Plain	Exp. cutoff	Devia-tion	Expon-ential
Erdös–Rényi [107]	√			√	√					√
PLRG [9], PLOD [221]	√			√	√		any γ (Eq. 8.5) (user-defined)			
Exponential cutoff [218]	√			√	√		any γ (Eq. 8.6) (user-defined)	√		
BA [37]	√						$\gamma = 3$			
Initial attractiveness [97]		√		√	√		$\gamma \in [2, \infty)$ (Eq. 9.3)			
AB [16]	√			√	√		$\gamma \in [2, \infty)$ (Eq. 9.4)			√
Edge Copying [179], [169]		√		√	$\gamma \in (1, \infty)$		√ (Eqs. 9.5, 9.6)			
GLP [62]	√			√	√		$\gamma \in (2, \infty)$ (Eq. 9.8)			
Accelerated growth [97], [38]	√			√	√		Power-law mixture of $\gamma = 2$ and $\gamma = 3$			
Fitness model [47]	√						$\gamma = 2.255$[1]			
Aiello et al. [10]		√					$\gamma \in [2, \infty)$ (Eq. 9.12)			
Pandurangan et al. [227]		√		√	$\gamma =?$		√			
Inet-3.0 [283]	√						$\gamma =?$[2]	√		
Forest Fire [191]		√					$\gamma =?$			
Pennock et al. [231]	√			√	√		$\gamma \in [2, \infty)$[3]		√	
Small-world [279]	√					√				√
Waxman [280]	√					√		√		
BRITE [205]	√					√	$\gamma =?$			
Yook et al. [286]	√					√	$\gamma =?$		√	
Fabrikant et al. [110]	√					√	$\gamma =?$			
RMat [73]	√	√	√	√	√		$\gamma =?$		√ (DGX)	

Table 8.2: *Taxonomy of graph generators (Contd.)*: The comparisons are made for graph diameter, existence of community structure (number of bipartite cores versus core size, or clustering coefficient $CC(k)$ of all nodes with degree k versus k), and clustering coefficient. N is the number of nodes in the graph. The empty cells represent information unknown to the authors, and require further research.

Generator	Diameter or Avg path len.	Community Bip. core vs size	Community C(k) vs k	Clustering coefficient	Remarks
Erdös–Rényi [107]	$O(\log N)$		Indep.	Low, $CC \propto N^{-1}$	
PLRG [9], PLOD [221]	$O(\log N)$	Indep.		$CC \to 0$ for large N	
Exponential cutoff [218]	$O(\log N)$			$CC \to 0$ for large N	
BA [37]	$O(\log N)$ or $O(\frac{\log N}{\log\log N})$			$CC \propto N^{-0.75}$	
Initial attractiveness [97]					
AB [16]					
Edge copying [169], [179]		Power law			
GLP [62]				Higher than AB, BA, PLRG	Internet only
Accelerated growth [99], [38]				Non-monotonic with N	
Fitness model [47]					
Aiello et al. [10]					
Pandurangan et al. [227]					
Inet [283]					Specific to the AS graph
Forest Fire [191]	"shrinks" as N grows				
Pennock et al. [231]					
Small-world [279]	$O(N)$ for small N, $O(\ln N)$ for large N, depends on p			$CC(p) \propto (1-p)^3$, Indep of N	N=num nodes p=rewiring prob
Waxman [280]					
BRITE [205]	Low (like in BA)			like in BA	BA + Waxman with additions
Yook et al. [286]					
Fabrikant et al. [110]					Tree, density 1
RMat [73]	Low (empirically)				

Table 8.3: *Table of detailed symbols*	
Symbol	**Description**
k	Random variable: degree of a node
$<k>$	Average degree of nodes in the graph
CC	Clustering coefficient of the graph
$CC(k)$	Clustering coefficient of degree-k nodes
γ	Power-law exponent: $y(x) \propto x^{-\gamma}$

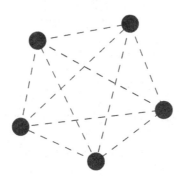

Figure 8.2: *The Erdös-Rényi model:* The black circles represent the nodes of the graph. Every possible edge occurs with equal probability.

For this reason, this model is often called the "Poisson" model.

Size of the largest component: Many properties of this model can be solved exactly in the limit of large N. A property is defined to hold for parameters (N, p) if the probability that the property holds on every graph in $G_{N,p}$ approaches 1 as $N \to \infty$. One of the most noted properties concerns the size of the largest component (subgraph) of the graph. For a low value of p, the graphs in $G_{N,p}$ have low density with few edges and all the components are small, having an exponential size distribution and finite mean size. However, with a high value of p, the graphs have a *giant component* with $O(N)$ of the nodes in the graph belonging to this component. The rest of the components again have an exponential size distribution with finite mean size. The changeover (called the *phase transition*) between these two regimes occurs at $p = \frac{1}{N}$. A heuristic argument for this is given below, and can be skipped by the reader.

Finding the phase transition point: Let the fraction of nodes not belonging to the giant component

---

[1] $P(k) \propto k^{-2.255} / \ln k$; [47] study a special case, but other values of the exponent γ may be possible with similar models.
[2] Inet-3.0 matches the Internet AS graph very well, but formal results on the degree-distribution are not available.
[3] $\gamma = 1 + \frac{1}{\alpha}$ as $k \to \infty$

be u. Thus, the probability of random node not belonging to the giant component is also u. But the neighbors of this node also do not belong to the giant component. If there are k neighbors, then the probability of this happening is u^k. Considering all degrees k, we get

$$
\begin{aligned}
u &= \sum_{k=0}^{\infty} p_k u^k \\
&= e^{-z} \sum_{k=0}^{\infty} \frac{(uz)^k}{k!} \quad \text{(using Eq 8.1)} \\
&= e^{-z} e^{uz} = e^{z(u-1)}
\end{aligned}
\tag{8.2}
$$

Thus, the fraction of nodes in the giant component is

$$
S = 1 - u = 1 - e^{-zS}
\tag{8.3}
$$

Equation 8.3 has no closed-form solutions, but we can see that when $z < 1$, the only solution is $S = 0$ (because $e^{-x} > 1 - x$ for $x \in (0, 1)$). When $z > 1$, we can have a solution for S, and this is the size of the giant component. The phase transition occurs at $z = p(N - 1) = 1$. Thus, a giant component appears only when p scales faster than N^{-1} as N increases.

Tree-shaped subgraphs: Similar results hold for the appearance of trees of different sizes in the graph. The critical probability at which almost every graph contains a subgraph of k nodes and l edges is achieved when p scales as N^z where $z = -\frac{k}{l}$ [49]. Thus, for $z < -\frac{3}{2}$, almost all graphs consist of isolated nodes and edges; when z passes through $-\frac{3}{2}$, trees of order 3 suddenly appear, and so on.

Diameter: Random graphs have a diameter concentrated around $\log N / \log z$, where z is the average degree of the nodes in the graph. Thus, the diameter grows slowly as the number of nodes increases.

Clustering coefficient: The probability that any two neighbors of a node are themselves connected is the connection probability $p = \frac{<k>}{N}$, where $< k >$ is the average node degree. Therefore, the clustering coefficient is:

$$
CC_{random} = p = \frac{< k >}{N}
\tag{8.4}
$$

Open questions and discussion: It is hard to exaggerate the importance of the Erdös-Rényi model in the development of modern graph theory. Even a simple graph-generation method has been shown to exhibit phase transitions and criticality. Many mathematical techniques for the analysis of graph properties were first developed for the random graph model.

However, even though random graphs exhibit such interesting phenomena, they do not match real-world graphs particularly well. Their degree distribution is Poisson (as shown by Equation 8.1), which has a very different shape from power laws or lognormals. There are no correlations between the degrees of adjacent nodes, nor does it show any form of "community" structure (which often shows up in real graphs like the WWW). Also, according to Equation 8.4, $\frac{CC_{random}}{<k>} = \frac{1}{N}$; but for many real-world graphs, $\frac{CC}{<k>}$ is independent of N (see Figure 9 from [17]).

Thus, even though the Erdös-Rényi random graph model has proven to be very useful in the early development of this field, it is not used in most of the recent work on modeling real graphs. To address some of these issues, researchers have extended the model to the so-called Generalized Random Graph Models, where the degree distribution can be set by the user (typically, set to be a power law).

8.1.2   GENERALIZED RANDOM GRAPH MODELS

Problem being solved: Erdös-Rényi graphs result in a Poisson degree distribution, which often conflicts with the degree distributions of many real-world graphs. Generalized random graph models extend the basic random graph model to allow arbitrary degree distributions.

Description and properties: Given a degree distribution, we can randomly assign a degree to each node of the graph so as to match the given distribution. Edges are formed by randomly linking two nodes till no node has extra degrees left. We describe two different models below: the PLRG model and the Exponential Cutoffs model. These differ only in the degree distributions used; the rest of the graph-generation process remains the same. The graphs thus created can, in general, include self-graphs and multigraphs (having multiple edges between two nodes).

The PLRG model: One of the obvious modifications to the Erdös-Rényi model is to change the degree distribution from Poisson to power law. One such model is the Power-Law Random Graph (PLRG) model of Aiello et al. [9] (a similar model is the *Power-Law Out Degree* (PLOD) model of Palmer and Steffan [221]). There are two parameters: α and β. The number of nodes of degree k is given by e^{α}/k^{β}.

PLRG degree distribution: By construction, the degree distribution is specifically a power law:

$$p_k \propto k^{-\beta} \tag{8.5}$$

where β is the power-law exponent.

PLRG connected component sizes: The authors show that graphs generated by this model can have several possible properties, based only on the value of β. When $\beta < 1$, the graph is almost surely connected. For $1 < \beta < 2$, a giant component exists, and smaller components are of size $O(1)$. For $2 < \beta < \beta_0 \sim 3.48$, the giant component exists and the smaller components are of size $O(\log N)$. At $\beta = \beta_0$, the smaller components are of size $O(\log N / \log \log N)$. For $\beta > \beta_0$, no giant component exists. Thus, for the giant component, we have a *phase transition* at $\beta = \beta_0 = 3.48$; there is also a change in the size of the smaller components at $\beta = 2$.

The Exponential cutoffs model: Another generalized random graph model is due to Newman et al. [218]. Here, the probability that a node has k edges is given by

$$p_k = Ck^{-\gamma}e^{-k/\kappa} \tag{8.6}$$

where C, γ, and κ are constants.

Exponential cutoffs degree distribution: This model has a power law (the $k^{-\gamma}$ term) augmented by an exponential cutoff (the $e^{-k/\kappa}$ term). The exponential cutoff, which is believed to be present in some social and biological networks, reduces the heavy-tail behavior of a pure power-law degree distribution. The results of this model agree with those of [9] when $\kappa \to \infty$.

Average path length for exponential cutoffs: Analytic expressions are known for the average path length of this model, but this typically tends to be somewhat less than that in real-world graphs [17].

Apart from PLRG and the exponential cutoffs model, some other related models have also been proposed. One important model is that of Aiello et al. [10], who assign weights to nodes and then form edges probabilistically based on the product of the weights of their end-points. The exact mechanics are, however, close to preferential attachment, and we discuss this later in Section 9.2.5.

Similar models have also been proposed for generating directed and bipartite random graphs. Recent work has provided analytical results for the sizes of the strongly connected components and cycles in such graphs [85, 99]. We do not discuss these any further; the interested reader is referred to [218].

Open questions and discussion: Generalized random graph models retain the simplicity and ease of analysis of the Erdös-Rényi model, while removing one of its weaknesses: the unrealistic Poisson degree distribution. However, most such models only attempt to match the degree distribution of real graphs, and no other patterns. For example, in most random graph models, the probability that two neighbors of a node are themselves connected goes as $O(N^{-1})$. This is exactly the clustering coefficient of the graph, and goes to zero for large N; but for many real-world graphs, $\frac{CC}{<k>}$ is independent of N (see Figure 9 from [17]). Also, many real world graphs (such as the WWW) exhibit the existence of communities of nodes, with stronger ties within the community than outside (see Section 16.2); random graphs do not appear to show any such behavior. Further work is needed to accommodate these patterns into the random graph generation process.

CHAPTER 9

Preferential Attachment and Variants

9.1 MAIN IDEAS

Problem being solved: Generalized random graph models try to model the power law or other degree distribution of real graphs. However, they do not make any statement about the *processes* generating the network. The search for a mechanism for network generation was a major factor in fueling the growth of the preferential attachment models, which we discuss below.

The rest of this section is organized as follows: in section 9.1.1, we describe the basic preferential attachment process. This has proven very successful in explaining many features of real-world graphs. Sections 9.1.3–9.2.7 describe progress on modifying the basic model to make it even more precise.

9.1.1 BASIC PREFERENTIAL ATTACHMENT

In the mid-1950s, Herbert Simon [249] showed that power-law tails arise when "the rich get richer." Derek Price applied this idea (which he called *cumulative advantage*) to the case of networks [91], as follows. We grow a network by adding vertices over time. Each vertex gets a certain out-degree, which may be different for different vertices but whose mean remains at a constant value m over time. Each outgoing edge from the new vertex connects to an old vertex with a probability proportional to the in-degree of the old vertex. This, however, leads to a problem since all nodes initially start off with in-degree zero. Price corrected this by adding a constant to the current in-degree of a node in the probability term, to get

$$P(\text{edge to existing vertex } v) = \frac{k(v) + k_0}{\sum_i (k(i) + k_0)}$$

where $k(i)$ represents the current in-degree of an existing node i, and k_0 is a constant.

A similar model was proposed by Barabási and Albert [37]. It has been a very influential model, and formed the basis for a large body of further work. Hence, we will look at the Barabási-Albert model (henceforth called the BA model) in detail.

Description of the BA model: The BA model proposes that structure emerges in network topologies as the result of two processes:

1. *Growth*: Contrary to several other existing models (such as random graph models) which keep a fixed number of nodes during the process of network formation, the BA model starts off with a small set of nodes and *grows* the network as nodes and edges are added over time.

2. *Preferential Attachment*: This is the same as the "rich get richer" idea. The probability of connecting to a node is proportional to the current degree of that node.

Using these principles, the BA model generates an *undirected* network as follows. The network starts with m_0 nodes, and grows in stages. In each stage, one node is added along with m edges which link the new node to m existing nodes (Figure 9.1). The probability of choosing an existing node as an endpoint for these edges is given by

$$P(\text{edge to existing vertex } v) = \frac{k(v)}{\sum_i k(i)} \tag{9.1}$$

where $k(i)$ is the degree of node i. Note that since the generated network is undirected, we do not need to distinguish between out-degrees and in-degrees. The effect of this equation is that nodes which already have more edges connecting to them get even more edges. This represents the "rich get richer" scenario.

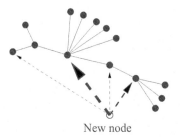

New node

Figure 9.1: *The Barabási–Albert model:* New nodes are added; each new node prefers to connect to existing nodes of high degree. The dashed lines show some possible edges for the new node, with thicker lines implying higher probability.

There are a few differences from Price's model. One is that the number of edges per new node is fixed at m (a positive integer); in Price's model only the mean number of added edges needed to be m. However, the major difference is that while Price's model generates a directed network, the BA model is undirected. This avoids the problem of the initial in-degree of nodes being zero; however, many real graphs are directed, and the BA model fails to model this important feature.

Properties of the BA model: We will now discuss some of the known properties of the BA model. These include the degree distribution, diameter, and correlations hidden in the model.

Degree distribution: The degree distribution of the BA model [98] is given by:

$$p_k \approx k^{-3}$$

for large k. In other words, the degree distribution has a power-law "tail" with exponent 3, independent of the value of m.

Diameter: Bollobás and Riordan [51] show that for large N, the diameter grows as $O(\log N)$ for $m = 1$, and as $O(\log N / \log \log N)$ for $m \geq 2$. Thus, this model displays the *small-world* effect: the distance between two nodes is, on average, far less than the total number of nodes in the graph.

Correlations between variables: Krapivsky and Redner [176] find two correlations in the BA model. First, they find that degree and age are positively correlated: older nodes have a higher mean degree. The second correlation is in the degrees of neighboring nodes, so that nodes with a similar degree are more likely to be connected. However, this asymptotically goes to 0 as $N \to \infty$.

Open questions and discussion: The twin ideas of *growth* and *preferential attachment* are definitely an immense contribution to the understanding of network generation processes. However, the BA model attempts to explain graph structure using *only* these two factors; most real-world graphs are probably generated by a slew of different factors. The price for this is some inflexibility in graph properties of the BA model.

- The power-law exponent of the degree distribution is fixed at $\gamma = 3$, and many real-world graphs deviate from this value.

- The BA model generates undirected graphs only; this prevents the model from being used for the many naturally occurring directed graphs.

- While Krapivsky and Redner show that the BA model should have correlations between node degree and node age (discussed above), Adamic and Huberman [3] apparently find no such correlations in the WWW.

- The generated graphs have exactly one connected component. However, many real graphs have several isolated components. For example, websites for companies often have private set of webpages for employees/projects only. These are a part of the WWW, but there are no paths to those webpages from outside the set. Military routers in the Internet router topology are another example.

- The BA model has a constant average degree of m; however, the average degree of some graphs (such as citation networks) actually increases over time according to a Densification Power Law [38, 99, 191] (see Section 3).

- The diameter of the BA model increases as N increases; however, many graphs exhibit shrinking diameters (see Section 3).

Also, further work is needed to confirm the existence or absence of a community structure in the generated graphs.

While the basic BA model does have these limitations, its simplicity and power make it an excellent base on which to build extended models. In fact, the bulk of graph generators in use today can probably trace their lineage back to this model. In the next few sections, we will look at some of these extensions and variations; as we will see, most of these are aimed at removing one or the other of the aforementioned limitations.

9.1.2   INITIAL ATTRACTIVENESS

Problem being solved: While the BA model generates graphs with a power-law degree distribution, the power-law exponent is stuck at $\gamma = 3$. Dorogovtsev et al. [97, 98] propose a simple one-parameter extension of the basic model which allows $\gamma \in [2, \infty)$. Other methods, such as the AB model described later, also do this, but they require more parameters.

Description and properties: The BA model is modified by adding an extra parameter $A \geq 0$ as follows:

$$P(\text{edge to existing vertex } v) = \frac{A + k(v)}{\sum_i (A + k(i))} \qquad (9.2)$$

where $k(i)$ is the degree of node i. The parameter A models the "initial attractiveness" of each site, and governs the probability of "young" sites gaining new edges. Note that the BA model is a special case of this model (when $A = 0$).

Degree distribution: The degree distribution is found to be a power law with exponent

$$\gamma = 2 + \frac{A}{m}$$

where m is the number of new edges being added at each timestep. Thus, depending on the value of A and m, $\gamma \in [2, \infty)$.

Open questions and discussion: This model adds a lot of flexibility to the BA model while requiring just a single parameter. As an extension of this, we could consider assigning different "initial attractiveness" values to different nodes; for example, this might be more realistic for new websites coming online on the WWW. Some progress has been made by Barabási and Bianconi [47], but their "fitness" parameters are used differently, and it is an open question what would happen if the parameter A in equation 9.2 were to be replaced by A_v.

9.1.3   INTERNAL EDGES AND REWIRING

Problem being solved: Graphs generated by the BA model have degree distributions with a power-law exponent of 3. However, the value of this exponent is often different for many naturally occurring graphs. The model described below attempts to remedy this.

Description and properties: In the BA model, one node and m edges are added to the graph every iteration. Albert and Barabási [16] decouple this addition of nodes and edges, and also extend the model by introducing the concept of edge rewiring. Starting with a small set of m_0 nodes, the resulting model (henceforth called the AB model) combines three processes:

- *With probability p, add m ($m \leq m_0$) new edges*: For each edge, one endpoint is chosen at random, and the other endpoint is chosen with probability

$$p(v) = \frac{k(v) + 1}{\sum_i (k(i) + 1)} \tag{9.3}$$

  where $p(v)$ represents the probability of node v being the endpoint, and $k(i)$ representing the degree of node i.

- *With probability q, rewire m links*: Choose a node i at random, and then choose one of its edges e_{ij}, remove it, and reconnect node i to some other node chosen using preferential attachment (Equation 9.3). This whole process is then repeated m times. This is effectively a way to locally reshuffle connections.

- *With probability $1 - p - q$, add a new node with m edges*: One end of these m edges is the new node; the other ends are chosen using preferential attachment (Equation 9.3). This was the only step in the BA model.

Note that in general, graphs generated by the AB model might have self-loops and multiple edges between two nodes.

Degree distribution: This model exhibits either a power-law or exponential degree distribution, depending on the parameters used. When $q < q_{max} = min(1 - p, (1 - p + m)/(1 + 2m))$, the distribution is a power law with exponent γ given by

$$\gamma = \frac{2m(1 - q) + 1 - p - q}{m} + 1 \tag{9.4}$$

However, for $q > q_{max}$, the distribution becomes exponential.

Validity of the model for the Internet graph: Chen et al. [78] try to check the validity of these processes in the context of the Internet. Their findings are summarized below:

- *Incremental Growth*: The Internet AS graph does grow incrementally, with nodes and edges being added gradually over time.

- *Linear Preferential Attachment*: However, they find that new AS's have a much stronger preference for connecting to the high-degree AS's than predicted by linear preferential attachment.

- *Addition of Internal Edges*: They also consider the addition of new edges between pre-existing AS's; this corresponds to the creation of new internal edges in the AB model. For the addition of every new internal edge, they put the end vertex with the smaller degree in a "Small Vertex" class, and the other end vertex in the "Large Vertex" class. They compare the degree distributions of these classes to that from the AS graph and find that while the "Small Vertex" class matches the real graph pretty well, the distribution of the "Large Vertex" class is very different between the AB model and the Internet.

- *Edge Rewiring*: They find that rewiring is probably not a factor in the evolution of the Internet.

Open questions and discussion: The AB model provides flexibility in the power-law exponent of the degree distribution. Further research is needed to show the presence or absence of a "community" structure in the generated graphs. Also, we are unaware of any work on analytically finding the diameter of graphs generated by this model.

9.2   RELATED METHODS

The initial idea of preferential attachment is powerful, illustrating that simple, local behavior leads to macroscopic properties (power-law tail in the degree distribution). Any other generator that explicitly or implicitly gives preference to high-degree nodes, will also lead to power laws. Next we list such generators.

9.2.1   EDGE COPYING MODELS

Problem being solved: Several graphs show community behavior, such as topic-based communities of websites on the WWW. Kleinberg et al. [169] and Kumar et al. [179] try to model this by using the intuition that most webpage creators will be familiar with webpages on topics of interest to them, and so when they create new webpages, they will link to some of these existing topical webpages. Thus, most new webpages will enhance the "topical community" effect of the WWW.

Description and properties: The Kleinberg [169] generator creates a directed graph. The model involves the following processes:

- *Node creation and deletion*: In each iteration, nodes may be independently created and deleted under some probability distribution. All edges incident on the deleted nodes are also removed.

- *Edge creation*: In each iteration, we choose some node v and some number of edges k to add to node v. With probability β, these k edges are linked to nodes chosen uniformly and independently at random. With probability $1 - \beta$, edges are *copied* from another node: we choose a node u at random, choose k of its edges (u, w), and create edges (v, w) (as shown in Figure 9.2). If the chosen node u does not have enough edges, all its edges are copied and the remaining edges are copied from another randomly chosen node.

- *Edge deletion*: Random edges can be picked and deleted according to some probability distribution.

This is similar to preferential attachment because the pages with high degree will be linked to by many other pages, and so have a greater chance of getting copied.

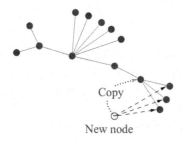

Copy

New node

Figure 9.2: *The edge copying model:* New nodes can choose to copy the edges of an existing node. This models the copying of links from other people's websites to create a new website.

Kumar et al. [179] propose a very similar model. However, there are some important differences. Whenever a new node is added, only *one* new edge is added. The edge need not be incident on the new node. With a probability α, the tail of the new edge (recall that this is a directed graph; the edge points from head to tail) is the new node, otherwise it is the tail of some randomly chosen existing edge. Similarly, with a probability β, the head of the new edge is the new node, otherwise it is the head of some random edge. Thus, the copying process takes place when head or tail of some existing edge gets chosen as the endpoint of the new edge.

Since important nodes on each "topic" might be expected to start off with a large number of edges incident on them, the edge copying process would tend to enhance the number of edges linking to them. Thus, the graph would gain several "communities," with nodes in the community linking to the "important" nodes of that community.

This and similar models have been analyzed by Kumar et al. [178]. They found the following interesting properties.

Degree distributions: For the Kleinberg model [169], the in-degree distribution is a power law with exponent given by

$$\gamma_{in} = \frac{1}{1 - \beta} \qquad (9.5)$$

For the model of Kumar et al. [179], both the in-degree and out-degree distributions follow power laws

$$\begin{aligned} \gamma_{in} &= \frac{1}{1 - \alpha} \\ \gamma_{out} &= \frac{1}{1 - \beta} \end{aligned} \qquad (9.6)$$

"Community" effect: They also show that such graphs can be expected to have a large number of bipartite cores (which leads to the community effect). However, more experiments might be needed to conclusively prove these results.

Open questions and discussion: The Kleinberg model [169] generates a tree; no "back-edges" are formed from the old nodes to the new nodes. Also, in the model of Kumar et al. [179], a fixed fraction of the nodes have zero in-degree or zero out-degree; this might not be the case for all real-world graphs (see Aiello et al. [10] for related issues).

However, the simple idea of copying edges can clearly lead to both power laws as well as community effects. "Edge copying" models are, thus, a very promising direction for future research.

9.2.2   MODIFYING THE PREFERENTIAL ATTACHMENT EQUATION

Problem being solved: Chen et al. [78] had found the AB model somewhat lacking in modeling the Web (Section 9.1.3). Specifically, they found that the preference for connecting to high-degree nodes is stronger than that predicted by linear preferential attachment. Bu and Towsley [62] attempt to address this issue.

Description and properties: The AB model [16] is changed by removing the edge rewiring process, and modifying the linear preferential attachment equation of the AB model to show higher preference for nodes with high degrees (as in [78]). Their new preferential attachment equation is:

$$p(v) = \frac{k(v) - \beta}{\sum_i (k(i) - \beta)} \qquad (9.7)$$

where $p(v)$ represents the probability of node v being the endpoint, $k(i)$ representing the degree of node i, and $\beta \in (-\infty, 1)$ is a tunable parameter. The smaller the value of β, the less preference is given to nodes with higher degree. Since $\beta < 1$, any node of degree 1 has a non-zero probability of acquiring new edges. This is called the GLP (Generalized Linear Preference) model.

Degree distribution: The degree distribution follows a power law with exponent

$$\gamma = \frac{2m - \beta(1 - p)}{(1 + p)m} + 1 \qquad (9.8)$$

Clustering coefficient: They also find empirically that the clustering coefficient for a GLP graph is much closer to that of the Internet than the BA, AB, and Power-Law Random Graph (PLRG [9]) models.

Bu and Towsley kept the preferential attachment equation linear (Equation 9.7); others such as Krapivsky and Redner [176] have studied *non-linear* preferential attachment:

$$p(v) \propto k^\alpha \qquad (9.9)$$

They achieve an important result, albeit a negative one. They find that power-law degree distributions occur only for linear attachment ($\alpha = 1$). When the preferential attachment is sublinear ($\alpha < 1$), the number of high-degree nodes decays faster than a power law. This goes against the findings of Chen et al. [78]. In the superlinear case ($\alpha > 1$), a single "gel" node emerges, which connects to nearly all other nodes. Again, many graphs, like the Internet, do not have this property.

9.2.3   MODELING INCREASING AVERAGE DEGREE

Problem being solved: The average degree of several real-world graphs (such as citation graphs) increases over time [38, 99, 191], according to a Densification Power Law. Barabási et al. [38] attempt to modify the basic BA model to accommodate this effect.

Description and properties: In their model, nodes join the graph at a constant rate, and form m edges to currently existing nodes with the linear preferential attachment equation (Equation 9.1), as in the BA model. *Also*, nodes already present in the graph form new internal edges, based on a different preferential attachment equation:

$$P(u, v) \propto k(u).k(v) \qquad (9.10)$$

In other words, the edge chooses *both* its endpoints by preferential attachment. The number of internal nodes added per iteration is proportional to the current number of nodes in the graph. Thus, it leads to the phenomenon of *accelerated growth*: the average degree of the graph increases linearly over time.

Degree distribution: However, the analysis of this model shows that it has two power-law regimes. The power-law exponent is $\gamma = 2$ for low degrees, and $\gamma = 3$ for high degrees. In fact, over a long period of time, the exponent converges to $\gamma = 2$.

Open questions and discussion: While this model allows for increasing average degree over time, the degree distribution is constrained to a power law with fixed exponents. Also, it is unknown if this model matches the "shrinking diameter" effect observed for growing graphs (see Section 3).

9.2.4   NODE FITNESS MEASURES

Problem being solved: The preferential attachment models noted above tend to have a correlation between the age of a node and its degree: higher the age, more the degree [176]. However, Adamic and Huberman find that this does not hold for the WWW [3]. There are websites which were created late but still have far higher in-degree than many older websites. Bianconi and Barabási [47] try to model this.

Description and properties: The model attaches a *fitness parameter* η_i to each node i, which does not change over time. The idea is that even a node which is added late could overtake older nodes in terms of degree, if the newer node has a much higher fitness value. The basic linear preferential attachment equation now becomes a *weighted* equation

$$P(\text{edge to existing vertex } v) = \frac{\eta_v k(v)}{\sum_i \eta_i k(i)} \tag{9.11}$$

Degree distribution: The authors analyze the case when the fitness parameters are drawn randomly from a uniform [0, 1] distribution. The resulting degree distribution is a power law with an extra inverse logarithmic factor. For the case where all fitness values are the same, this model becomes the simple BA model.

Open questions and discussion: Having a node's popularity depend on its "fitness" intuitively makes a lot of sense. Further research is needed to determine the distribution of node fitness values in real-world graphs. For this "fitness distribution," we also need to compute the corresponding degree distribution, and ensure that it matches reality.

9.2.5   GENERALIZING PREFERENTIAL ATTACHMENT

Problem being solved: The BA model is undirected. A simple adaptation to the directed case is: new edges are created to point from the new nodes to existing nodes chosen preferentially according to their *in-degree*. However, the out-degree distribution of this model would not be a power law. Aiello et al. [10] propose a very general model for generating directed graphs which give power laws for both in-degree and out-degree distributions. A similar model was also proposed by Bollobás et al. [50].

Description and properties: The basic idea is the following:

- Generate four random numbers $m(n, n)$, $m(n, e)$, $m(e, n)$, and $m(e, e)$ according to some bounded probability distributions; the numbers need not be independent.

- One node is added to the graph in each iteration.

- $m(n, n)$ edges are added from new node to new node (forming self-loops).

- $m(n, e)$ edges are added from the new node to random existing nodes, chosen preferentially according to their in-degree (higher in-degree nodes having higher chance of being chosen).

- $m(e, n)$ edges are added from existing nodes to the new node; the existing nodes are chosen randomly with probability proportional to their out-degrees.

- $m(e, e)$ edges are added between existing nodes. Again, the choices are proportional to the in-degrees and out-degrees of the nodes.

Finally, nodes with no in- and out-degrees are ignored.

Degree distributions: The authors show that even in this general case, both the in-degree and out-degree distributions follow power laws, with the following exponents:

$$
\begin{aligned}
\gamma_{in} &= 2 + \frac{m(n, n) + m(e, n)}{m(n, e) + m(e, e)} \\
\gamma_{out} &= 2 + \frac{m(n, n) + m(n, e)}{m(e, n) + m(e, e)}
\end{aligned}
\tag{9.12}
$$

A similar result is obtained by Cooper and Frieze [84] for a model which also allows some edge endpoints to be chosen uniformly at random, instead of always via preferential attachment.

Open questions and discussion: The work referenced above shows that even a very general version of preferential attachment can lead to power-law degree distributions. Further research is needed to test for all the other graph patterns, such as diameter, community effects, and so on.

9.2.6 PAGERANK-BASED PREFERENTIAL ATTACHMENT

Problem being solved: Pandurangan et al. [227] found that the *PageRank* [59] values for a snapshot of the Web graph follow a power law. They propose a model that tries to match this *PageRank* distribution of real-world graphs, *in addition to* the degree distributions.

Description and properties: They modify the basic preferential attachment mechanism by adding a *PageRank*-based preferential attachment component:

- With probability a, new edges preferentially connect to higher-degree nodes. This is typical preferential attachment.

- With probability b, new edges preferentially connect to nodes with high *PageRank*. According to the authors, this represents linking to nodes which are found by using a *search engine* which uses PageRank-based rankings.

- With probability $1 - a - b$, new edges connect to randomly chosen nodes.

Degree and PageRank distributions: They empirically show that this model can match both the degree distributions as well as the *PageRank* distribution of the Web graph. However, closed-form formulas for the degree distributions are not provided for this model.

Open questions and discussion: This model offers an intuitive method of incorporating the effects of Web search engines into the growth of the Web. However, the authors also found that the plain edge-copying model of Kumar et al. [179] could *also* match the *PageRank* distribution (in addition to the degree distributions) without specifically attempting to do so. Thus, this work might be taken to be another alternative model of the Web.

9.2.7   THE FOREST FIRE MODEL

Problem being solved: Leskovec et al. [191] develop a preferential-attachment based model which matches the Densification Power Law and the shrinking diameter patterns of graph evolution, in addition to the power-law degree distribution.

Description and properties: The model has two parameters: a *forward burning probability* p, and a *backward burning ratio* r. The graph grows one node at a time. The new node v adds links to the existing nodes according to a "forest fire" process:

1. *Pick ambassador:* Node v chooses an *ambassador* node w uniformly at random, and links to w.

2. *Select some of the ambassador's edges:* A random number x is chosen from a binomial distribution with mean $(1 - p)^{-1}$. Node v then selects x edges of w, both in-links and out-links, but selecting in-links with probability r times less than out-links. Let $w_1, w_2, \ldots, w_x$ be the other ends of these selected edges.

3. *Follow these edges and repeat:* Node v forms edges pointing to each of these nodes $w_1, \ldots, w_x$, and then recursively applies step (2) to these nodes.

This process conceptually captures a "forest fire" in the existing graph; the fire starts at the ambassador node and then probabilistically spreads to the other nodes if they are connected to nodes which are currently "burning." Some nodes end up creating large "conflagrations," which form many out-links before the fire dies out, thus resulting in power laws.

Degree distributions: Both the in-degree and out-degree distribution are empirically found to follow power laws.

Community structure: This method is similar to the edge copying model discussed earlier (Section 9.2.1) because existing links are "copied" to the new node v as the fire spreads. This leads to a community of nodes, which share similar edges.

Densification Power Law and Shrinking Diameter: The Forest Fire model empirically seems to follow both of these patterns. The intuition behind densification is clear: as the graph grows, the chances of a larger fire also grow, and so new nodes have higher chances of getting more edges. However, the intuition behind the shrinking diameter effect is not clear.

Open questions and discussion: This is certainly a very interesting and intuitive model, but the authors note that rigorous analysis of this model appears to be quite difficult. The *RMat* generator (discussed later in section 11) and its recently proposed generalization into *Kronecker graphs* [190] is one possible approach that offers formal results for these graph patterns.

9.3   SUMMARY OF PREFERENTIAL ATTACHMENT MODELS

All preferential attachment models use the idea that the "rich get richer:" high-degree nodes attract more edges, or high-PageRank nodes attract more edges, and so on. This simple process, along with the idea of network growth over time, *automatically* leads to the power-law degree distributions seen in many real-world graphs. As such, these models made a very important contribution to the field of graph mining. Still, most of these models appear to suffer from some limitations: for example, they do not seem to generate any "community" structure in the graphs they generate. Also, apart from the work of Pennock et al. [231], little effort has gone into finding reasons for deviations from power-law behaviors for some graphs. It appears that we need to consider additional processes to understand and model such characteristics.

CHAPTER 10

Incorporating Geographical Information

Both the random graph and preferential attachment models have neglected one attribute of many real graphs: the constraints of geography. For example, it is easier (cheaper) to link two routers which are physically close to each other; most of our social contacts are people we meet often, and who consequently probably live close to us (say, in the same town or city); and so on. In the following paragraphs, we discuss some important models which try to incorporate this information.

10.1 EARLY MODELS

10.1.1 THE SMALL-WORLD MODEL

Problem being solved: The small-world model is motivated by the observation that most real-world graphs seem to have low average distance between nodes (a global property), but have high clustering coefficients (a local property). Two experiments from the field of sociology shed light on this phenomenon.

Travers and Milgram [268] conducted an experiment where participants had to reach randomly chosen individuals in the U.S.A. using a chain letter between close acquaintances. Their surprising find was that, for the chains that completed, the average length of the chain was only six, in spite of the large population of individuals in the "social network." While only around 29% of the chains were completed, the idea of small paths in large graphs was still a landmark find.

The reason behind the short paths was discovered by Mark Granovetter [133], who tried to find out how people found jobs. The expectation was that the job seeker and his eventual employer would be linked by long paths; however, the actual paths were empirically found to be very short, usually of length one or two. This corresponds to the low average path length mentioned above. Also, when asked whether a friend had told them about their current job, a frequent answer of the respondents was *"Not a friend, an acquaintance"*. Thus, this low average path length was being caused by acquaintances, with whom the subjects only shared *weak ties*. Each acquaintance belonged to a different social circle and had access to different information. Thus, while the social graph has a high clustering coefficient (i.e., is "clique-ish"), the low diameter is caused by weak ties joining faraway cliques.

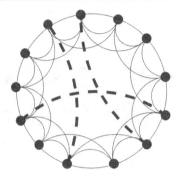

Figure 10.1: *The small-world model:* Nodes are arranged in a ring lattice; each node has links to its immediate neighbors (solid lines) and some long-range connections (dashed lines).

Description and properties: Watts and Strogatz [279] independently came up with a model which had exactly these characteristics: it has a *high clustering coefficient* but *low diameter*. Their model (Figure 10.1), which has only one parameter p, is described below.

- *Regular ring lattice:* Start with a ring lattice (N, k): this is a graph with N nodes set in a circle. Each node has k edges to its closest neighbors, with $k/2$ edges on each side. This is the set of *close friendships*, and has a high clustering coefficient. Let $N \gg k \gg \ln N \gg 1$.

- *Rewiring:* For each node u, each of its edges (u, v) is rewired with probability p to form some different edge (u, w), where node w is chosen uniformly at random. Self-loops and duplicate edges are forbidden. This accounts for the *weak acquaintances*.

Distance between nodes, and clustering coefficient: With $p = 0$, the graph remains a plain ring lattice. Both the clustering coefficient and the average distance between nodes are fairly high ($CC(p = 0) \sim 3/4$ and $L(p = 0) \sim N/2k \gg 1$). Thus, small-world structure is absent. When $p = 1$, both the clustering coefficient and the average distance are fairly low ($CC(p = 1) \sim k/N \ll 1$ and $L(p = 1) \sim \ln N / \ln k$). Thus, the graph is not "clique-ish" enough. However, there exists a range of p values for which $L(p) \sim L(1)$ but $CC(p) \gg CC(1)$; that is, the average distance remains low while the clustering coefficient is high. These are exactly the desired properties.

The reason for this is that the introduction of a few long-range edges (which are exactly the weak ties of Granovetter) leads to a highly nonlinear effect on the average distance L. Distance is contracted not only between the endpoints of the edge, but also their immediate neighborhoods (circles of friends). However, these few edges lead to a very small change in the clustering coefficient. Thus, we get a broad range of p for which the small-world phenomenon coexists with a high clustering coefficient.

Degree distribution: All nodes start off with degree k, and the only changes to their degrees are due to rewiring. The shape of the degree distribution is similar to that of a random graph, with a strong peak at k, and it decays exponentially for large k.

Open questions and discussion: The small-world model is very successful in combining two important graph patterns: small diameters and high clustering coefficients. However, the degree distribution decays exponentially, and does not match the power-law distributions of many real-world graphs. Extension of the basic model to power-law distributions is a promising research direction.

10.1.2 THE WAXMAN MODEL

Problem being solved: The Internet graph is constrained by geography: it is cheaper to link two routers which are close in distance. Waxman [280] proposed a very simple model which focuses on this interaction of network generation with geography (Figure 10.2).

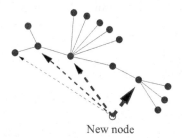

New node

Figure 10.2: *The Waxman model:* New nodes prefer to connect to existing nodes which are closer in distance.

Description and properties: The Waxman generator places random points in Cartesian two-dimensional space (representing the placement of routers on land). An edge (u, v) is placed between two points u and v with probability

$$P(u, v) = \beta \exp \frac{-d(u, v)}{L\alpha} \qquad (10.1)$$

Here, α and β are parameters in the range $(0, 1)$, $d(u, v)$ is the Euclidean distance between points u and v, and L is the maximum Euclidean distance between points.

The parameters α and β control the geographical constraints. The value of β affects the *edge density:* larger values of β result in graphs with higher edge densities. The value of α relates the short edges to longer ones: a small value of α increases the density of short edges relative to longer edges.

Open questions and discussion: The Waxman generator has been very popular in the networking community. However, it does not yield a power law-degree distribution, and further work is needed to analyze the other graph patterns for this generator.

10.1.3   THE BRITE GENERATOR

Problem being solved: Medina et al. [205] try to combine the geographical properties of the Waxman generator with the incremental growth and preferential attachment techniques of the BA model. Their graph generator, called BRITE, has been extensively used in the networking community for simulating the structure of the Internet.

Description and properties: The main features of BRITE are:

- *Node placement*: The geography is assumed to be a square grid. Nodes can either be placed randomly, or with a heavy-tailed distribution.

- *Links per node*: As in the BA model, this is set to m, a model parameter.

- *Incremental Growth*: Either we could start off by placing all the nodes and then adding links (as in the Waxman model), or we could add nodes and links as we go along (as in the BA model). The latter gives us incremental growth.

- *Wiring of edges*: The authors provide three cases: (1) The edges could link randomly chosen nodes. (2) We could have pure preferential connectivity, as in the BA model. (3) The interesting case is when we combine preferential connectivity with geographical constraints. Suppose that we want to add an edge to node u. The probability of the other endpoint of the edge being node v is a *weighted* preferential attachment equation, with the weights being the probability of that edge existing in the pure Waxman model (Equation 10.1)

$$P(u, v) = \frac{w(u, v)k(v)}{\sum_i w(u, i)k(i)} \tag{10.2}$$
$$\text{where } w(u, v) = \beta \exp \frac{-d(u, v)}{L\alpha} \text{ as in Eq. 10.1}$$

Open questions and discussion: The emphasis of BRITE is on creating a system that can be used to generate different kinds of topologies. This allows the user a lot of flexibility, and is one reason behind the widespread use of BRITE in the networking community. However, there is little discussion of parameter fitting: how can the model parameters be set so as to generate synthetic graphs which successfully match the properties (such as the power-law exponent) of some given real-world graph? Developing algorithms for parameter fitting, and understanding the scenarios which lead to power-law graphs (such as the *Heuristically Optimized Tradeoffs* model described later in Section 10.2.2) are interesting avenues for further research.

10.1.4 OTHER GEOGRAPHICAL CONSTRAINTS

Problem being solved: Yook et al. [286] find two interesting linkages between geography and networks (specifically the Internet):

1. The geographical distribution of Internet routers and Autonomous Systems (AS) is a fractal, and is strongly correlated with population density. This is intuitive: more people require more bandwidth and more routers. This finding is at odds with most of the previous models, which usually expect the nodes to be spread uniformly at random in some geographical area (BRITE allows inhomogeneous distributions, but not fractals).

2. They plot the probability of an edge versus the length of the edge and find that the probability is *inversely proportional* to the Euclidean distance between the endpoints of the edge. They explain this by saying that the cost of linking two routers is essentially the cost of administration (fixed) and the cost of the physical wire (proportional to distance). For long links, the distance-cost dominates, so the probability of the link should be inversely proportional to distance. However, in the Waxman and BRITE models, this probability decays exponentially with length (Equation 10.1).

Description and properties: To remedy the first problem, they suggest using a self-similar geographical distribution of nodes. For the second problem, they propose a modified version of the BA model. Each new node u is placed on the map using the self-similar distribution, and adds edges to m existing nodes. For each of these edges, the probability of choosing node v as the endpoint is given by a modified preferential attachment equation:

$$P(\text{node } u \text{ links to existing node } v) \propto \frac{k(v)^{\alpha}}{d(u,v)^{\sigma}} \tag{10.3}$$

where $k(v)$ is the current degree of node v and $d(u,v)$ is the Euclidean distance between the two nodes. The values α and σ are parameters, with $\alpha = \sigma = 1$ giving the best fits to the Internet. They show that varying the values of α and σ can lead to significant differences in the topology of the generated graph.

Similar geographical constraints may hold for social networks as well: individuals are more likely to have friends in the same city as compared to other cities, in the same state as compared to other states, and so on recursively. Watts et al. [278] and (independently) Kleinberg [167] propose a hierarchical model to explain this phenomenon; we will discuss both in more detail in Section 20.3.

10.2   TOPOLOGY FROM RESOURCE OPTIMIZATIONS

Most of the methods described above have approached power-law degree distributions from the preferential-attachment viewpoint: if the "rich get richer," power laws might result. However, another point of view is that power laws can result from *resource optimizations*. We will discuss some such models below.

10.2.1   THE HIGHLY OPTIMIZED TOLERANCE MODEL

Problem being solved: Carlson and Doyle [66, 100] have proposed an optimization-based reason for the existence of power laws in graphs. They say that power laws may arise in systems due to *tradeoffs* between yield (or profit), resources (to prevent a risk from causing damage), and tolerance to risks.

Description and properties: As an example, suppose we have a forest which is prone to forest fires. Each portion of the forest has a different chance of starting the fire (say, the dryer parts of the forest are more likely to catch fire). We wish to minimize the damage by assigning resources such as firebreaks at different positions in the forest. However, the total available resources are limited. The problem is to place the firebreaks so that the expected cost of forest fires is minimized.

In this model, called the *Highly Optimized Tolerance* (HOT) model, we have n possible events (starting position of a forest fire), each with an associated probability $p_i (1 \leq i \leq n)$ (dryer areas have higher probability). Each event can lead to some *loss* l_i, which is a function of the resources r_i allocated for that event: $l_i = f(r_i)$. Also, the total resources are limited: $\sum_i r_i \leq R$ for some given R. The aim is to minimize the expected cost

$$J = \left\{ \sum_i p_i l_i \mid l_i = f(r_i), \sum_i r_i \leq R \right\} \qquad (10.4)$$

Degree distribution: The authors show that if we assume that cost and resource usage are related by a power law $l_i \propto r_i^\beta$, then, under certain assumptions on the probability distribution p_i, resources are spent on places having higher probability of costly events. In fact, resource placement is related to the probability distribution p_i by a power law. Also, the probability of events which cause a loss greater than some value k is related to k by a power law.

The salient points of this model are:

- high efficiency, performance, and robustness to designed-for uncertainties

- hypersensitivity to design flaws and unanticipated perturbations

- nongeneric, specialized, structured configurations, and

- power laws.

Resilience under attack: This concurs with other research regarding the vulnerability of the Internet to attacks. Several researchers have found that while a large number of randomly chosen nodes and edges can be removed from the Internet graph without appreciable disruption in service, attacks *targeting* important nodes can disrupt the network very quickly and dramatically [20, 220]. The HOT model also predicts a similar behavior: since routers and links are *expected* to be down occasionally, it is a "designed-for" uncertainty and the Internet is impervious to it. However, a *targeted* attack is not designed for, and can be devastating.

Newman et al. [217] modify HOT using a utility function which can be used to incorporate "risk aversion." Their model (called *Constrained Optimization with Limited Deviations* or COLD) truncates the tails of the power laws, lowering the probability of disastrous events.

HOT has been used to model the sizes of files found on the WWW. The idea is that dividing a single file into several smaller files leads to faster load times, but increases the cost of navigating through the links. They show good matches with this dataset.

Open questions and discussion: The HOT model offers a completely new recipe for generating power laws; power laws can result as a by-product of resource optimizations. However, this model requires that the resources be spread in an *globally-optimal* fashion, which does not appear to be true for several large graphs (such as the WWW). This led to an alternative model by Fabrikant et al. [110], which we discuss below.

10.2.2 THE HEURISTICALLY OPTIMIZED TRADEOFFS MODEL

Problem being solved: The previous model requires globally-optimal resource allocations. However, graphs like the Internet appear to have evolved by *local* decisions taken by the engineers/administrators on the spot. Fabrikant et al. [110] propose an alternative model in which the graph grows as a result of trade-offs made *heuristically* and locally (as opposed to optimally, for the HOT model).

Description and properties: The model assumes that nodes are spread out over a geographical area. One new node is added in every iteration, and is connected to the rest of the network with *one* link. The other endpoint of this link is chosen to optimize between two conflicting goals: (1) minimizing the "last-mile" distance, that is, the *geographical* length of wire needed to connect a new node to a pre-existing graph (like the Internet), and (2) minimizing the transmission delays based on number of hops, or, the distance along the network to reach other nodes. The authors try to optimize a linear combination of the two (Figure 10.3). Thus, a new node i should be connected to an existing node j chosen to minimize

$$\alpha.d_{ij} + h_j \ (j < i) \tag{10.5}$$

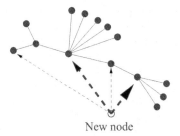

New node

Figure 10.3: *The Heuristically Optimized Tradeoffs Model:* A new node prefers to link to existing nodes which are both close in distance and occupy a "central" position in the network.

where d_{ij} is the distance between nodes i and j, h_j is some measure of the "centrality" of node j, and α is a constant that controls the relative importance of the two.

Degree distribution: The authors find that the characteristics of the network depend greatly on the value of α: when α is less than a particular constant based on the shape of the geography, the "centrality" constraints dominate and the generated network is a *star* (one central "hub" node which connects to all the other "spoke" nodes). On the other hand, when α grows as fast as $\log N$, the geographical constraints dominate and the degree distribution falls off exponentially fast. However, if α is anywhere in between, power-law degree distributions result.

Open questions and discussion: As in the *Highly Optimized Tolerance* model described before (Section 10.2.1), power laws are seen to fall off as a by-product of resource optimizations. However, only local optimizations are now needed, instead of global optimizations. This makes the *Heuristically Optimized Tradeoffs* model very appealing.

In its current version, however, the model only generates graphs of density 1 (that is, one edge per node). This also implies that the graph is actually a tree, whereas many real-world graphs have cycles (for example, a node might have multiple connections to the Internet to maintain connectivity in case one of its links fails). Also, in addition to the degree distribution, the generated graphs need to be analyzed for all other graph patterns too. Further research needs to modify the basic model to address these issues. One step in this direction is the recent work of Berger et al. [44], who generalize the *Heuristically Optimized Tradeoffs* model, and show that it is equivalent to a form of preferential attachment; thus, competition between opposing forces can give rise to preferential attachment, and we already know that preferential attachment can, in turn, lead to power laws and exponential cutoffs.

10.3   GENERATORS FOR THE INTERNET TOPOLOGY

While the generators described above are applicable to any graphs, some special-purpose generators have been proposed to specifically model the Internet topology. Structural generators exploit
the hierarchical structure of the Internet, while the Inet generator modifies the basic preferential
attachment model to better fit the Internet topology. We look at both of these below.

10.3.1   STRUCTURAL GENERATORS

Problem being solved: Work done in the networking community on the structure of the Internet
has led to the discovery of *hierarchies* in the topology. At the lowest level are the Local Area Networks
(LANs); a group of LANs are connected by *stub domains*, and a set of *transit domains* connect the
stubs and allow the flow of traffic between nodes from different stubs. More details are provided in
Section 5.1. However, the previous models do not explicitly enforce such hierarchies on the generated
graphs.

Description and properties: Calvert et al. [65] propose a graph generation algorithm which specifically models this hierarchical structure. The general topology of a graph is specified by six parameters,
which are the numbers of transit domains, stub domains, and LANs, and the number of nodes in
each. More parameters are needed to model the connectivities within and across these hierarchies.
To generate a graph, points in a plane are used to represent the locations of the centers of the transit
domains. The nodes for each of these domains are spread out around these centers, and are connected
by edges. Now, the stub domains are placed on the plane and are connected to the corresponding
transit node. The process is repeated with nodes representing LANs.

   The authors provide two implementations of this idea. The first, called *Transit-Stub*, does not
model LANs. Also, the method of generating connected subgraphs is to keep generating graphs till
we get one that is connected. The second, called *Tiers*, allows multiple stubs and LANs, but allows
only one transit domain. The graph is made connected by connecting nodes using a minimum
spanning tree algorithm.

Open questions and discussion: These models can specifically match the hierarchical nature of
the Internet, but they make no attempt to match any other graph pattern. For example, the degree
distributions of the generated graphs need not be power laws. Hence, while these models have been
widely used in the networking community, they need modifications to be as useful in other settings.

   Tangmunarunkit et al. [259] compare such structural generators against generators which
focus only on power-law distributions. They find that even though power-law generators do not
explicitly model hierarchies, the graphs generated by them have a substantial level of hierarchy,
though not as strict as with the generators described above. Thus, the hierarchical nature of the
structural generators can also be mimicked by other generators.

10.3.2 THE INET TOPOLOGY GENERATOR

Problem being solved: Winick and Jamin [283] developed the Inet generator to model only the Internet Autonomous System (AS) topology, and to match features specific to it.

Description and properties: Inet-2.2 generates the graph by the following steps:

- Each node is assigned a degree from a power-law distribution with an exponential cutoff (as in Equation 8.6).

- A spanning tree is formed from all nodes with a degree greater than 1.

- All nodes with degree one are attached to this spanning tree using linear preferential attachment.

- All nodes in the spanning tree get extra edges using linear preferential attachment till they reach their assigned degree.

The main advantage of this technique is in ensuring that the final graph remains connected.

However, under this scheme, too many of the low-degree nodes get attached to other low-degree nodes. For example, in the Inet-2.2 topology, 35% of degree-2 nodes have adjacent nodes with degree 3 or less; for the Internet, this happens only for 5% of the degree-2 nodes. Also, the highest degree nodes in Inet-2.2 do not connect to as many low-degree nodes as the Internet. To correct this, Winick and Jamin came up with the Inet-3 generator, with a modified preferential attachment system.

The preferential attachment equation now has a weighting factor which uses the degrees of the nodes on both ends of some edge. The probability of a degree i node connecting to a degree j node is

$$P(\text{degree } i \text{ node connects to degree } j \text{ node}) \propto w_i^j . j \qquad (10.6)$$

$$\text{where } w_i^j = MAX\left(1, \sqrt{\left(\log\frac{i}{j}\right)^2 + \left(\log\frac{f(i)}{f(j)}\right)^2}\right) \qquad (10.7)$$

Here, $f(i)$ and $f(j)$ are the number of nodes with degrees i and j respectively, and can be easily obtained from the degree distribution equation. Intuitively, what this weighting scheme is doing is the following: when the degrees i and j are close, the preferential attachment equation remains linear. However, when there is a large difference in degrees, the weight is the Euclidean distance between the points on the log-log plot of the degree distribution corresponding to degrees i and j, and this distance increases with increasing difference in degrees. Thus, edges connecting nodes with a big difference in degrees are preferred.

Open questions and discussion: Inet has been extensively used in the networking literature.

However, the fact that it is so specific to the Internet AS topology makes it somewhat unsuitable for any other topologies.

10.4 COMPARISON STUDIES

While a large body of work has been done on developing new graph generators, effort has also gone into comparing different graph generators, especially on certain graphs like the Internet. However, different studies have used different metrics for comparing different graph generators. We combine the results of several studies in our discussion below [20, 62, 259, 260].

Now we will describe the results for each of these metrics. When possible, we will try to explain the reasons behind the results; we note, however, that most of these results are as yet formally unexplained.

Distortion: Consider any spanning tree T on the graph. Then, the *distortion for T* is the average distance on T between any two nodes that are connected by an edge in the graph. The *distortion for the graph* is the smallest such average over all possible spanning trees.

The graph distortion measures the difference between the real graph and its spanning tree. Tangmunarunkit et al. [259, 260] use heuristics to evaluate this metric, and find that both the Internet AS level and Router level graphs have low distortion, and the PLRG model matches this. The Waxman model (section 10.1.2) has high distortion. Inet-3.0 is found to show similar distortion as the Internet AS level graph [283].

Expansion: Both the Internet AS level and Router level graphs exhibit high expansion[1] [259, 260]. The PLRG model (section 8.1.2) matches this pattern. The Tiers model (section 10.3.1) has low expansion.

Resilience under random failures: Both the Internet AS level and Router level graphs show high resilience under random failures [220, 259, 260]. The PLRG and AB (section 9.1.3) models match this: in fact, power-law graphs remain unaffected even when as many as 5% of the nodes are randomly chosen and removed [20, 50]. However, for graphs with exponentially decaying degree distributions, such as the Erdós-Rényi random graph (Section 8.1.1) and the Small-World model (Section 10.1.1), the average diameter increases monotonically as random nodes are removed. The Transit-Stub model (Section 10.3.1) also has low resilience.

Resilience under targeted attacks: When nodes are removed in decreasing order of degree, the situation is the complete reverse of the "random failures" scenario. Power-law graphs show drastic increases in average diameter (doubling the original value as the top 5% of the highest-degree nodes are removed),

---

[1]For the expansion, resilience and distortion metrics, Tangmunarunkit et al. [259, 260] distinguish between only two states: "low" and "high" (for example, exponential expansion is high, and sub-exponential is low).

while exponential graphs exhibit more or less the same behavior as for the *failure* case. This brings to the forefront the importance of the most-connected nodes in power-law graphs [20, 50, 220].

Hierarchical structure of the Internet: Even though power-law models do not lead to any explicit hierarchy of nodes, a hierarchy shows up nonetheless in graphs generated by such models. Tangmunarunkit et al. [259, 260] surmise that this hierarchy is a side effect of the power-law distribution: nodes with high degree function appear to be near the top of the hierarchy, while low-degree nodes form the leaves.

Characteristic path length: The AB, GLP, and Inet models (Sections 9.1.3, 9.2.2, and 10.3.2) give similar path lengths as the Internet AS level graph with proper choice of parameters, while PLRG does not. PLRG also shows very high variance for this metric [62]. Inet-3.0 has similar characteristic path lengths over time as the Internet AS graph [283].

Clustering coefficient: The clustering coefficient of GLP is closer to the Internet AS level graph than those for AB, PLRG, and Inet [62]. Inet-3.0 exhibits lower clustering coefficient than the Internet AS graph, possibly because it does not have a large, dense clique connecting the high-degree "core" nodes, as is seen in the Internet graph [283].

In addition, Winick and Jamin [283] compared the Inet-3.0 generator to the Internet AS graph for several other patterns, and observed good fits for many of these. Note, however, that Inet-3.0 was developed specifically for the Internet AS topology.

Reasons behind the resilience properties of the Internet: Much effort has gone into understanding the resilience properties of the Internet (resilient under random failures, but drastically affected by targeted attacks). Albert et al. [20] propose that in power-law graphs like the Internet, the high-degree nodes are the ones "maintaining" most of the connectivity, and since there are so few of them, it is unlikely for them to be chosen for removal under random failures. Targeted attacks remove exactly these nodes, leading to severe connectivity failures.

Tauro et al. [261] propose a solution based on the structure of the Internet AS graph. They say that the graph is organized as a *Jellyfish* (or concentric rings around a core), with the most important nodes in the core, and layers further from the core decreasing in importance (see Section 5.1). Random node removal mostly removes one-degree nodes which hang off the core or the layers, and do not affect connectivity. However, targeted node removal removes nodes from the (small) core and then successively from the important layers; since most nodes connect in or toward the central core, this leads to a devastating loss of connectivity. Perhaps the Internet power-law graph was generated in a fashion such that the "core" nodes achieved the highest connectivity; that would agree with both [261] and [20].

Interestingly, similar behavior is exhibited by metabolic pathways in organisms; Jeong et al. [148] show that the diameter does not change under random node removal, but increases fourfold when only 8% of the most connected nodes are removed. Solé and Montoya [250] see the same thing with ecological networks, and Newman et al. [216] for email networks.

CHAPTER 11

The *RMat* (Recursive MATrix) Graph Generator

We have seen that most of the current graph generators focus on only one graph pattern – typically the degree distribution – and give low importance to all the others. There is also the question of how to fit model parameters to match a given graph. What we would like is a tradeoff between parsimony (few model parameters), realism (matching most graph patterns, if not all), and efficiency (in parameter fitting and graph generation speed). In this section, we present the *RMat* generator, which attempts to address all of these concerns.

Problem being solved: The *RMat* [73] generator tries to meet several desiderata:

- The generated graph should match several graph patterns, including *but not limited to* power-law degree distributions (such as hop-plots and eigenvalue plots).

- It should be able to generate graphs exhibiting deviations from power laws, as observed in some real-world graphs [231].

- It should exhibit a strong "community" effect.

- It should be able to generate directed, undirected, bipartite, or weighted graphs with the same methodology.

- It should use as few parameters as possible.

- There should be a fast parameter-fitting algorithm.

- The generation algorithm should be efficient and scalable.

Description and properties:

The *RMat* generator creates directed graphs with 2^n nodes and E edges, where both values are provided by the user. We start with an empty adjacency matrix, and divide it into four equal-sized partitions. One of the four partitions is chosen with probabilities a, b, c, d, respectively ($a + b + c + d = 1$), as in Figure 11.1. The chosen partition is again subdivided into four smaller partitions, and the procedure is repeated until we reach a simple cell (=1 × 1 partition). The nodes (that is, row and column) corresponding to this cell are linked by an edge in the graph. This process is repeated

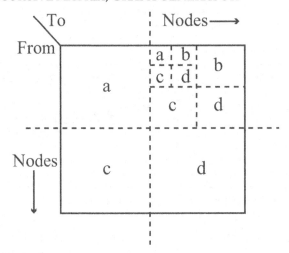

Figure 11.1: *The RMat model:* The adjacency matrix is broken into four equal-sized partitions, and one of those four is chosen according to a (possibly non-uniform) probability distribution. This partition is then split recursively till we reach a single cell, where an edge is placed. Multiple such edge placements are used to generate the full synthetic graph.

E times to generate the full graph. There is a subtle point here: we may have *duplicate* edges (i.e., edges which fall into the same cell in the adjacency matrix), but we only keep one of them when generating an unweighted graph. To smooth out fluctuations in the degree distributions, some noise is added to the (a, b, c, d) values at each stage of the recursion, followed by renormalization (so that $a + b + c + d = 1$). Typically, $a \geq b, a \geq c, a \geq d$.

Degree distribution: There are only three parameters (the partition probabilities a, b, and c; $d = 1 - a - b - c$). The skew in these parameters ($a \geq d$) leads to lognormals and the DGX [46] distribution, which can successfully model both power-law and "unimodal" distributions [231] under different parameter settings.

<u>*Communities:*</u> Intuitively, this technique is generating "communities" in the graph:

- The partitions a and d represent separate groups of nodes which correspond to communities (say, "Linux" and "Windows" users).

- The partitions b and c are the *cross-links* between these two groups; edges there would denote friends with separate preferences.

- The recursive nature of the partitions means that we automatically get sub-communities within existing communities (say, "RedHat" and "Mandrake" enthusiasts within the "Linux" group).

Diameter, singular values and other properties: We show experimentally that graphs generated by *RMat* have small diameter and match several other criteria as well.

Extensions to undirected, bipartite, and weighted graphs: The basic model generates directed graphs; all the other types of graphs can be easily generated by minor modifications of the model. For undirected graphs, a directed graph is generated and then made symmetric. For bipartite graphs, the same approach is used; the only difference is that the adjacency matrix is now rectangular instead of square. For weighted graphs, the number of *duplicate* edges in each cell of the adjacency matrix is taken to be the weight of that edge. More details may be found in [73].

Parameter fitting algorithm: We are given some input graph, and need to fit the *RMat* model parameters so that the generated graph matches the input graph in terms of graph patterns.

We can calculate the expected degree distribution: the probability p_k of a node having out-degree k is given by

$$p_k = \frac{1}{2^n}\binom{E}{k}\sum_{i=0}^{n}\binom{n}{i}\left[\alpha^{n-i}(1-\alpha)^i\right]^k\left[1-\alpha^{n-i}(1-\alpha)^i\right]^{E-k}$$

where 2^n is the number of nodes in the *RMat* graph, E is the number of edges, and $\alpha = a + b$. Fitting this to the out-degree distribution of the input graph provides an estimate for $\alpha = a + b$. Similarly, the in-degree distribution of the input graph gives us the value of $b + c$. Conjecturing that the $a : b$ and $a : c$ ratios are approximately $75 : 25$ (as seen in many real world scenarios), we can calculate the parameters (a, b, c, d).

Next, we show experimentally that *RMat* can match both power-law distributions as well as deviations from power laws.

Experiments: We show experiments on the following graphs (see Table 1.2):
Epinions: A directed graph of who-trusts-whom from epinions.com [239]: $N = 75,879$; $E = 508,960$.
Epinions-U: An undirected version of the *Epinions* graph: $N = 75,879$; $E = 811,602$.
Clickstream: A bipartite graph of Internet users' browsing behavior [211]. An edge (u, p) denotes that user u accessed page p. It has 23,396 users, 199,308 pages, and 952,580 edges.

The graph patterns we look at are:

1. Both in-degree and out-degree distributions.

2. "Hop-plot" and "effective diameter:" The "hop-plot" shows the number of reachable pairs of nodes, versus the number of hops (see Definitions 2.3 and 2.4).

3. Singular value vs. rank plots: Singular values [235] are similar to eigenvalues (they are the same for undirected graphs), but eigenvalues may not exist for bipartite graphs, while singular values do.

4. "Singular vector value" versus rank plots: The "singular vector value" of a node is the absolute value of the corresponding component of the first singular vector of the graph. It can be considered to be a measure of the "importance" of the node, and as we will see later, is closely related to the widely used concept of "Bonacich centrality" in social network analysis.

5. "Stress" distribution: The "stress" of an edge is the number of shortest paths between node pairs that it is a part of (see Definition 16.2).

In addition to *RMat*, we show the fits achieved by some other models, chosen for their popularity or recency: these are the *AB*, *GLP*, and *PG* models (Sections 9.1.3, 9.2.2, and [231] respectively). All three can only generate undirected graphs; thus, we can compare them with *RMat* only on *Epinions-U*. The parameters of these three models are set by exhaustive search; we use the terms *AB+*, *PG+*, and *GLP+* to stand for the original models augmented by our parameter fitting.

Epinions-U: Figure 11.2 shows the comparison plots on this undirected graph. *RMat* gives the closest fits. Also, note that all the y-scales are logarithmic, so small differences in the plots actually represent significant deviations.

Epinions: Figure 11.3 shows results on this directed graph. The *RMat* fit is very good; the other models considered are not applicable.

Clickstream: Figure 11.4 shows results on this bipartite graph. As before, the *RMat* fit is very good. In particular, note that the in-degree distribution is a power law while the out-degree distribution deviates significantly from a power law; *RMat* matches *both* of these very well. This is because *RMat* generates a "truncated discrete lognormal" (a DGX distribution [46]) which, under the correct parameter settings, can give good fits to power laws as well. Again, the other models are not applicable.

Open questions and discussion: While the *RMat* model shows promise, there has not been any thorough analytical study of this model. Also, it seems that only three parameters might not provide enough degrees of freedom to match all varieties of graphs; a step in this direction is the *Kronecker graph generator* [190], which generalizes the *RMat* model and can match several patterns such as the Densification Power Law and the shrinking diameters effect (see Section 3) in addition to all the patterns that *RMat* matches.

As we mentioned earlier, *RMat* is the basis behind the so-called *graph500* generator[1] which is very well supported, and heavily used for supercomputing benchmarks.

---

[1]http://www.graph500.org/

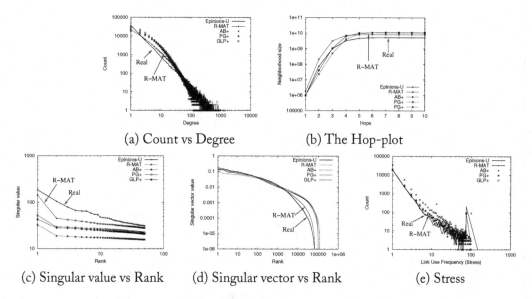

(a) Count vs Degree　　　　(b) The Hop-plot

(c) Singular value vs Rank　　(d) Singular vector vs Rank　　(e) Stress

Figure 11.2: *Epinions-U undirected graph:* We show (a) degree, (b) hop-plot, (c) singular value, (d) "singular vector value," and (e) stress distributions for the *Epinions-U* dataset. *RMat* gives the best matches to the *Epinions-U* graph, among all the generators. In fact, for the stress distribution, the *RMat* and *Epinions-U* plots are almost indistinguishable.

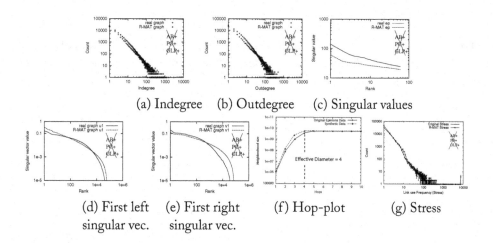

(a) Indegree　　(b) Outdegree　　(c) Singular values

(d) First left　　(e) First right　　(f) Hop-plot　　(g) Stress
singular vec.　　singular vec.

Figure 11.3: *Epinions directed graph:* The *AB+*, *PG+*, and *GLP+* methods **do not apply**. The crosses and dashed lines represent the *RMat* generated graphs, while the pluses and strong lines represent the real graph.

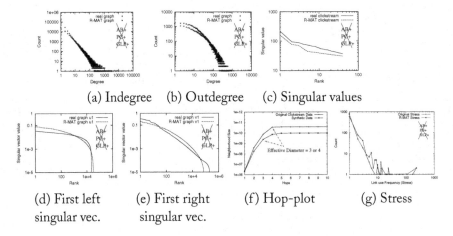

(a) Indegree   (b) Outdegree   (c) Singular values

(d) First left   (e) First right   (f) Hop-plot   (g) Stress
singular vec.   singular vec.

Figure 11.4: *Clickstream bipartite graph:* The *AB+*, *PG+*, and *GLP+* methods **do not apply**. The crosses and dashed lines represent the *RMat* generated graphs, while the pluses and strong lines represent the real graph.

CHAPTER 12

Graph Generation by Kronecker Multiplication

The *RMat* generator described in the previous paragraphs achieves its power mainly via a form of recursion: the adjacency matrix is recursively split into equal-sized quadrants over which edges are distributed unequally. One way to generalize this idea is via Kronecker matrix multiplication, wherein one small initial matrix is recursively "multiplied" with itself to yield large graph topologies. The mathematical simplicity of this generative model yields simple closed-form expressions for several measures of interest, such as degree distributions and diameters, thus enabling ease of analysis and parameter-fitting.

Problem being solved: The *RMat* model, while often yielding good matches for several real-world patterns, does not lend itself easily to mathematical analysis of these patterns. Graph generation via Kronecker multiplication not only generalizes *RMat*, but also leads to simple closed-form formulas for several patterns of interest.

Description and properties: We first recall the definition of the Kronecker product.

Definition 12.1 Kronecker product of matrices Given two matrices $\mathbf{A} = [a_{i,j}]$ and $\mathbf{B}$ of sizes $n \times m$ and $n' \times m'$ respectively, the Kronecker product matrix $\mathbf{C}$ of dimensions $(n * n') \times (m * m')$ is given by

$$\mathbf{C} = \mathbf{A} \otimes \mathbf{B} \doteq \begin{pmatrix} a_{1,1}\mathbf{B} & a_{1,2}\mathbf{B} & \dots & a_{1,m}\mathbf{B} \\ a_{2,1}\mathbf{B} & a_{2,2}\mathbf{B} & \dots & a_{2,m}\mathbf{B} \\ \vdots & \vdots & \ddots & \vdots \\ a_{n,1}\mathbf{B} & a_{n,2}\mathbf{B} & \dots & a_{n,m}\mathbf{B} \end{pmatrix} \tag{12.1}$$

In other words, for any nodes X_i and X_j in $\mathbf{A}$ and X_k and X_ℓ in $\mathbf{B}$, we have nodes $X_{i,k}$ and $X_{j,\ell}$ in the Kronecker product $\mathbf{C}$, and an edge connects them iff the edges (X_i, X_j) and (X_k, X_ℓ) exist in $\mathbf{A}$ and $\mathbf{B}$. The Kronecker product of two graphs is the Kronecker product of their adjacency matrices.

Let us consider an example. Figure 12.1(a–c) shows the recursive construction of $G \otimes H$, when $G = H$ is a three node path. Consider node $X_{1,2}$ in Figure 12.1(c): It belongs to the H graph that replaced node X_1 (see Figure 12.1(b)), and in fact is the X_2 node (i.e., the center) within this small H-graph. Thus, the graph H is recursively embedded "inside" graph G.

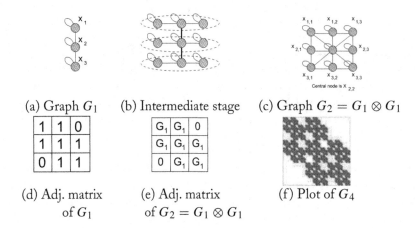

| (a) Graph G_1 | (b) Intermediate stage | (c) Graph $G_2 = G_1 \otimes G_1$ |

1	1	0
1	1	1
0	1	1

G_1	G_1	0
G_1	G_1	G_1
0	G_1	G_1

(d) Adj. matrix
of G_1

(e) Adj. matrix
of $G_2 = G_1 \otimes G_1$

(f) Plot of G_4

Figure 12.1: *Example of Kronecker multiplication:* Top: a "three chain" and its Kronecker product with itself; each of the X_i nodes gets expanded into three nodes, which are then linked together. Bottom row: the corresponding adjacency matrices, along with matrix for the fourth Kronecker power G_4.

The Kronecker graph generator simply applies the Kronecker product multiple times over. Starting with a binary *initiator* graph $\mathcal{G}_1$ with N_1 nodes and E_1 edges, successively larger graphs are produced by repeated Kronecker multiplication: $\mathcal{G}_2 = \mathcal{G}_1 \otimes \mathcal{G}_1, \mathcal{G}_3 = \mathcal{G}_2 \otimes \mathcal{G}_1, \ldots, \mathcal{G}_n = \mathcal{G}_{n-1} \otimes \mathcal{G}_1$. The recursive nature of the generator is obvious. It is also clear that the properties of $\mathcal{G}_n$ depend on those of the initiator graph $\mathcal{G}_1$. The exact form of these properties is discussed below.

Degree distribution: Kronecker graphs have multinomial degree distributions, for both in- and out-degrees. This follows from the fact that the degrees of nodes in $\mathcal{G}_k$ can be expressed as the kth Kronecker power of the vector of node degrees of the initiator graph $\mathcal{G}_1$.

Diameter: If the initiator graph $\mathcal{G}_1$ has a diameter of d, and each node has an edge to itself (a self-loop), then the diameter of any generated Kronecker graph $\mathcal{G}_k$ is d, irrespective of k.

A similar statement can also be made about the *effective diameter* – recall that the effective diameter is the minimum number of hops in which some fraction q (say, $q = 0.9$) of all connected pairs of nodes can reach each other (see Definition 2.4). If $\mathcal{G}_1$ has diameter d and a self-loop on every node, then for every q, the effective diameter of $\mathcal{G}_k$ using fraction q of reachable pairs converges to d (from below) as k increases.

Eigenvalues and eigenvectors: Just as with node degrees, Kronecker graphs have multinomial distribution of eigenvalues. Similarly, the components of each eigenvector also follow a multinomial distribution.

Community structure: The self-similar nature of the Kronecker graph product is clear: To produce $\mathcal{G}_k$ from $\mathcal{G}_{k-1}$, we "expand" (replace) each node of $\mathcal{G}_{k-1}$ by converting it into a copy of $\mathcal{G}_1$, and we join these copies together according to the adjacencies in $\mathcal{G}_{k-1}$ (see Figure 12.1). This process is very natural: one can imagine it as positing that communities with the graph grow recursively, with nodes in the community recursively getting expanded into miniature copies of the community. Nodes in the subcommunity then link among themselves and also to nodes from different communities.

Densification Power Law: If the initiator graph $\mathcal{G}_1$ has N_1 nodes and E_1 edges, then the series of graphs generated by Kronecker multiplication follow the Densification Power Law (DPL) with densification exponent $a = \log(E_1)/\log(N_1)$.

To summarize, thanks to its simple mathematical structure, Kronecker graph generation allows the derivation of closed-form formulas for several important patterns. Of particular importance are the "temporal" patterns regarding changes in properties as the graph grows over time: both the constant diameter and the densification power-law patterns are similar to those observed in real-world graphs [191], and are not matched by most graph generators.

Open questions and discussion: While Kronecker multiplication allows several patterns to be computed analytically, its discrete nature leads to "staircase effects" in the degree and spectral distributions. A modification of the aforementioned generator avoids these effects, as follows. Instead of the initiator matrix $\mathcal{G}_1$ being binary, its elements are allowed to be *probabilities*. Thus, all graphs $\mathcal{G}_k$ have probabilities in each cell as well. Specific graph adjacency matrices can now be generated by filling in each cell with 0 or 1 depending on the cell probability. The authors find graphs generated in this fashion to avoid the staircase effect while giving good matches to common patterns, but the analysis of such graphs is still an open question.

More recent variations include the tensor-based model RTM [15], the random typing graph generator RTG [13], and the random dot product model [287]. For a formal analysis of Kronecker graphs, see [162].

CHAPTER 13

Summary and Practitioner's Guide

We have seen many graph generators in the preceding pages. Is any generator the "best?" Which one should we use?

The answer seems to depend on the application area: the *Inet* generator is specific to the Internet and can match its properties very well, the *BRITE* generator allows geographical considerations to be taken into account, "edge copying" models provide a good intuitive mechanism for modeling the growth of the Web along with matching degree distributions and community effects, and so on.

However, recent activity seems to focus on *RMat*, the basis behind the *graph500* generator[1] which is attracting high interest from the research community. Quoting the *graph500* website:

> Backed by a steering committee of over 50 international HPC experts from academia, industry, and national laboratories, Graph 500 will establish a set of large-scale benchmarks for these [*graph*] applications. The Graph 500 steering committee is in the process of developing comprehensive benchmarks to address three application kernels: concurrent search, optimization (single source shortest path), and edge-oriented (maximal independent set). Further, we are in the process of addressing five graph-related business areas: Cybersecurity, Medical Informatics, Data Enrichment, Social Networks, and Symbolic Networks.

The *graph500* site also offers open source code for graph generation, version 2.1.4 at the time of writing, as well as the specifications for benchmarks (version 1.2). It is designed to generate graphs from "toy" size (10^{10}bytes $\approx$17GB) to "huge" size (10^{15}bytes $\approx$ 1.1PB). According to the site, *RMat* and *graph500* have been used in three international competitions, with international participants, including industry (IBM), government (LBL, LANL, ORNL), and universities (Georgia Tech, Moscow State Univ.). The competing machines included top-of-the-line machines, like "IBM BlueGene" and "Jaguar."

We expect that *RMat* was chosen by the *graph500* committee for two reasons: it offers (a) several desirable properties (thanks to its recursive, fractal-like construction) and (b) very easy parallelization.

---

[1]http://www.graph500.org/

PART III

Tools and Case Studies

CHAPTER 14

SVD, Random Walks, and Tensors

Here we review some powerful tools and, specifically, the singular value decomposition (SVD) of a matrix, its uses for ranking in web queries, and its extension to multi-mode matrices ("tensors"). Table 14.1 gives a list of symbols and the definitions we use.

To help with the intuition, we shall often resort to an $N \times n$ matrix $\mathbf{A}$, which will stand, e.g., for who-bought-what: We assume N users and n products, with matrix $\mathbf{A} = [a_{i,j}]$ reporting how many units of product j were bought by customer i. Alternatively, it can stand for a document-term matrix, where each document is represented in the vector-space model as a set of words.

Questions of interest are, e.g., are there clusters of customers (or products), or similarly, which pairs of products sell together.

Symbol	Definition
$\mathbf{A}$	matrix (bold capital)
$\mathbf{A}^t$	transpose of matrix $\mathbf{A}$
$a_{i,j}$	i, j element of matrix $\mathbf{A}$
$\vec{p}$	a column vector $[p_1, \ldots, p_n]^t$
p_i	the i-th element of vector $\vec{p}$
c	scalar (lower case)
$\mathcal{X}$	a tensor (calligraphic font)
$x_{i,j,k}$	the element of $\mathcal{X}$ at i-th row, j-th column, k-th tube

Figure 14.1: Symbols and definitions. Matrices are in bold-capital font, scalars are in lower-case, normal font, vectors have an arrow superscript, and tensors are in calligraphic font.

14.1 EIGENVALUES—DEFINITION AND INTUITION

Arbitrary, non-symmetric matrices   For any square matrix $\mathbf{T}$, we have:

Definition 14.1   For a square matrix $\mathbf{T}$, a scalar λ, and a vector $\vec{u}$ are called an eigenvalue-eigenvector pair if

$$\mathbf{T}\vec{u} = \lambda\vec{u}$$

That is, the vector $\vec{u}$ remains parallel to itself, after the rotation and scaling that the matrix $\mathbf{T}$ implements. λ is the scaling that vector $\vec{u}$ undergoes. Figure 14.2 illustrates the geometric intuition: for the matrix

$$\mathbf{T} = \begin{bmatrix} 3 & 1.3 \\ 1.3 & 2 \end{bmatrix}$$

the figure plots the results of multiplication with $\mathbf{T}$ of several unit vectors (forming the unit circle in the figure). Notice that the tips of the arrows form an ellipse whose properties are closely related to the eigen-decomposition of matrix $\mathbf{T}$:

- the major axis of the ellipse corresponds to the first eigenvector ($\vec{u}_1 = [0.82, 0.56]$ here)
- the minor axis is perpendicular to the major, and corresponds to the second eigenvector ($\vec{u}_2 = [-0.56, 0.82]$)
- the length of the major axis is the first eigenvalue ($\lambda_1 = 3.89$ in our case) and similarly for the minor axis ($\lambda_2 = 1.11$)

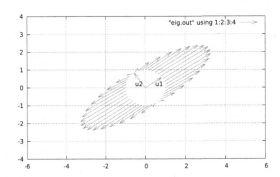

Figure 14.2: Geometric illustration of eigenvectors: points on the unit circle are translated to points on the ellipse periphery, when multiplied with the $\mathbf{T}$ matrix of the example. The eigenvectors $\vec{u}_1$ and $\vec{u}_2$ are shown, corresponding to the axis of the ellipse.

For non-symmetric matrices, the eigenvalues and eigenvectors may be complex numbers. We will not study non-symmetric matrices further, but we will only mention the powerful Perron-Frobenius theorem, which holds for a class of *non-negative* matrices[1], the so-called *irreducible* ones. This theorem is the reason that the upcoming PageRank algorithm works (see Section 14.4).

Definition 14.2 Irreducible matrix   A matrix $\mathbf{T}$ is *irreducible* if it is square, non-negative, and every node i can reach every node j – formally, for every i and j, there is a k such that $\mathbf{T}^k(i, j) > 0$.

---

[1] A matrix is non-negative if all its elements are non-negative

That is, there exists a number of steps k, such that any node i can reach any node j, after k steps. $\mathbf{T}^k$ is the k-th power of matrix $\mathbf{T}$, and $\mathbf{T}(i, j)$ is the element in the i-th row and j-th column.

Then we have:

Theorem 14.3 Perron-Frobenius *If* $\mathbf{T}$ *is an irreducible matrix, then it has a real eigenvalue* λ_{max} *such that other eigenvalues satisfy* $|\lambda| < \lambda_{max}$.

Proof See [253], p. 177.   **QED**

Symmetric matrices   Building on the intuition for the customer-product matrix $\mathbf{A}$ above, let's consider the $n \times n$ product-to-product similarity matrix $\mathbf{T}$, which we define as $\mathbf{A}^t \times \mathbf{A}$.

Let $\mathbf{T}$ be a symmetric $n \times n$ matrix of rank r ($r < n$). Then we have (e.g., see [262], p. 3):

Definition 14.4   A symmetric matrix $\mathbf{T}$ has the eigen-decomposition

$$\mathbf{T} = \mathbf{U}\Sigma\mathbf{U}^t$$

with $\mathbf{U}$ being orthonormal ($\mathbf{U} \times \mathbf{U}^t = \mathbf{U}^t \times \mathbf{U} = \mathbf{I}_r$) and Σ is diagonal.

$\mathbf{I}_r$ is the $r \times r$ identity matrix. All the elements of Σ are real numbers.

As we see later, there is a close relationship between the eigenvalue decomposition of the product-product similarity matrix $\mathbf{T} = \mathbf{A}^t\mathbf{A}$ and the singular value decomposition of the customer-product matrix $\mathbf{A}$.

14.2   SINGULAR VALUE DECOMPOSITION (SVD).

Motivation: Given a who-bought-what matrix of N users and n products, how can we find groups and trends? SVD answers exactly this question.

Intuition: As we saw, the eigenvalues and eigenvectors are defined for square matrices. For rectangular matrices, a closely related concept is the Singular Value Decomposition (SVD) [131, 234, 254].

term document	data	information	retrieval	brain	lung
CS-TR1	1	1	1	0	0
CS-TR2	2	2	2	0	0
CS-TR3	1	1	1	0	0
CS-TR4	5	5	5	0	0
MED-TR1	0	0	0	2	2
MED-TR2	0	0	0	3	3
MED-TR3	0	0	0	1	1

Consider a set of N points ('customers') represented as a $N \times n$ matrix $\mathbf{A}$, as in Table 14.2. Such a matrix could represent, e.g., N patients with n numerical symptoms each (blood pressure, cholesterol level, etc.), or N sales with n products in a data mining application [8], with the dollar amount spent on each product, by the given sale, etc. For concreteness, we shall assume that it represents N documents (rows) with n terms (columns) each, as in the "bag-of-words" (= vectors space model) of Information Retrieval [242], [103]. It would be desirable to group similar documents together, as well as similar terms together. This is exactly what SVD does, automatically! In fact, SVD creates a *linear* combination of terms, as opposed to non-linear ones that, e.g., Kohonen's neural networks could provide [193, 240]. Nevertheless, these groups of terms are valuable: in Information Retrieval terminology, each would correspond to a "concept."

Formulas: The formal definition for SVD follows:

Theorem 14.5   SVD   *Given an $N \times n$ real matrix $\mathbf{A}$ we can express it as*

$$\mathbf{A} = \mathbf{U} \times \Lambda \times \mathbf{V}^t \tag{14.1}$$

where $\mathbf{U}$ is a column-orthonormal $N \times r$ matrix, r is the rank of the matrix $\mathbf{A}$, Λ is a diagonal $r \times r$ matrix, and $\mathbf{V}$ is a column-orthonormal $k \times r$ matrix.

Proof: See [235, p. 59].   **QED**

The entries of Λ are non-negative. If we insist that the diagonal matrix Λ has its elements sorted in descending order, then the decomposition is *unique*[2].

Recall that a matrix $\mathbf{U}$ is column-orthonormal iff its column vectors are mutually orthogonal and of unit length. Equivalently:

$$\mathbf{U}^t \times \mathbf{U} = \mathbf{I} \tag{14.2}$$

where $\mathbf{I}$ is the identity matrix. Schematically, see Figure 14.3.

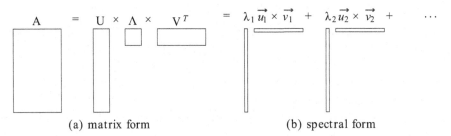

(a) matrix form   (b) spectral form

Figure 14.3: Illustration of SVD, in matrix form and spectral form.

Eq. 14.1 equivalently states that a matrix $\mathbf{A}$ can be brought in the form

$$\mathbf{A} = \lambda_1 \vec{u}_1 \times (\vec{v}_1)^t + \lambda_2 \vec{u}_2 \times (\vec{v}_2)^t + \ldots + \lambda_r \vec{u}_r \times (\vec{v}_r)^t \tag{14.3}$$

[2]Except when there are equal entries in Λ, in which case they and their corresponding columns of $\mathbf{U}$ and $\mathbf{V}$ can be permuted.

where $\vec{u}_i$, and $\vec{v}_i$ are column vectors of the **U** and **V** matrices respectively, and λ_i the diagonal elements of the matrix Λ. Intuitively, the SVD identifies "rectangular blobs" of related values in the **A** matrix. For example, for the above "toy" matrix of Table 14.2, we have two "blobs" of values, while the rest of the entries are zero. This is confirmed by the SVD, which identifies them both:

$$
\mathbf{A} = 9.64 \times \begin{bmatrix} 0.18 \\ 0.36 \\ 0.18 \\ 0.90 \\ 0 \\ 0 \\ 0 \\ 0 \\ 0 \\ 0 \\ 0 \\ 0.53 \\ 0.80 \\ 0.27 \end{bmatrix} \times [0.58,\ 0.58,\ 0.58,\ 0,\ 0] +
$$

$$
5.29 \times \begin{bmatrix} 0 \\ 0.53 \\ 0.80 \\ 0.27 \end{bmatrix} \times [0,\ 0,\ 0,\ 0.71,\ 0.71]
$$

or

$$
\mathbf{A} = \begin{bmatrix} 0.18 & 0 \\ 0.36 & 0 \\ 0.18 & 0 \\ 0.90 & 0 \\ 0. & 0.53 \\ 0. & 0.80 \\ 0. & 0.27 \end{bmatrix} \times \begin{bmatrix} 9.64 & 0 \\ 0 & 5.29 \end{bmatrix} \times \begin{bmatrix} 0.58 & 0.58 & 0.58 & 0 & 0 \\ 0 & 0 & 0 & 0.71 & 0.71 \end{bmatrix}
$$

Notice that the rank of the matrix is $r=2$: there are effectively two types of documents (CS and medical documents) and two "concepts," i.e., groups-of-terms: the "CS concept" (that is, the group {"data," "information," "retrieval"}), and the "medical concept" (that is, the group {"lung," "brain"}). The intuitive meaning of the **U** and **V** matrices is as follows: **U** can be thought of as the *document-to-concept* similarity matrix, while **V**, symmetrically, is the *term-to-concept* similarity matrix. For example, $v_{1,2} = 0$ means that the first term ("data") has zero similarity with the second concept (the "lung-brain" concept).

Properties, and relation to eigenvalues. The SVD is a powerful operation, with several applications. We list some observations, which are useful for graph mining, multimedia indexing, and Information Retrieval.

Observation 14.6   The $N \times N$ matrix $\mathbf{D} = \mathbf{A} \times \mathbf{A}^t$ will intuitively give the document-to-document similarities – in our arithmetic example, it is

$$\mathbf{D} = \mathbf{A} \times \mathbf{A}^t = \begin{bmatrix} 3 & 6 & 3 & 15 & 0 & 0 & 0 \\ 6 & 12 & 6 & 30 & 0 & 0 & 0 \\ 3 & 6 & 3 & 15 & 0 & 0 & 0 \\ 15 & 30 & 15 & 75 & 0 & 0 & 0 \\ 0 & 0 & 0 & 0 & 8 & 12 & 4 \\ 0 & 0 & 0 & 0 & 12 & 18 & 6 \\ 0 & 0 & 0 & 0 & 4 & 6 & 2 \end{bmatrix}$$

Matrix $\mathbf{D}$ is a *Gramian* (or *Gram* or *positive semi-definite*) matrix, that is, it is the product of a matrix with its transpose, i.e., $\mathbf{D} = \mathbf{A} \times \mathbf{A}^t$. Such matrices have non-negative eigenvalues. Moreover:

Lemma 14.7   *For Gramian matrices, the SVD and the eigen-decomposition coincide.*

Proof See [262] p. 6.   **QED**

The symmetric observations hold for the transposed case:

Observation 14.8   Symmetrically, the $n \times n$ matrix $\mathbf{T} = \mathbf{A}^t \times \mathbf{A}$ will give the term-to-term similarities – in our example, it is:

$$\mathbf{T} = \mathbf{A}^t \times \mathbf{A} = \begin{bmatrix} 31 & 31 & 31 & 0 & 0 \\ 31 & 31 & 31 & 0 & 0 \\ 31 & 31 & 31 & 0 & 0 \\ 0 & 0 & 0 & 14 & 14 \\ 0 & 0 & 0 & 14 & 14 \end{bmatrix}$$

Lemma 14.9   *Both $\mathbf{D}$ and $\mathbf{T}$ have the same eigenvalues, which are the squares of the singular values of $\mathbf{A}$ (that is, the λ_i elements of the Λ matrix of the SVD of $\mathbf{A}$).*

Proof All the above observations can be proved from Theorem 14.5 and from the fact that $\mathbf{U}$ and $\mathbf{V}$ are column-orthonormal:

$$\mathbf{A} \times \mathbf{A}^t = \mathbf{U} \times \Lambda \times \mathbf{V}^t \times \mathbf{V} \times \Lambda \times \mathbf{U}^t = \mathbf{U} \times \Lambda \times \mathbf{I} \times \Lambda \times \mathbf{U}^t = \mathbf{U} \times \Lambda^2 \times \mathbf{U}^t$$

and similarly for $\mathbf{A}^t \times \mathbf{A}$.   **QED**

Corollary 14.10   *The eigenvectors of the $\mathbf{D}$ ($\mathbf{T}$) matrix will be the columns of the $\mathbf{U}$ ($\mathbf{V}$) matrix of the SVD of $\mathbf{A}$.*

Note that the SVD is extremely useful for several settings that involve least-squares optimization, such as in regression, in under-constraint and over-constraint linear problems, etc. See [235] or [254] for more details. Next, we show how it has been applied for Information Retrieval and filtering, under the name of (LSI).

Discussion: SVD is one of the many matrix decomposition methods, also known as matrix factorization methods. It is extremely popular, with implementations available in all major mathematical and statistical packages (matlab^{TM}, mathematica^{TM}, octave, R). Some of the most successful alternatives are the following:

- Independent Component Analysis (ICA) [145][223]. ICA is also known as "blind source separation," and strives to achieve sparse encoding. Intuitively, it also gives soft-clustering, but each member belongs to few groups, much fewer than with SVD.
- Semi-Discrete Decomposition (SDD), where the values in the U and V matrices have to take integer values (-1,0,1), and the Λ is identity. This has advantages in information retrieval settings [171].
- Non-Negative Matrix Factorization (NNMF): SVD may give negative entries in the U and V matrices. When envisioned as soft-clustering and cluster membership, negative scores are hard to interpret. NNMF guarantees that all scores are non-negative. The original one is [182] with several recent variations like a sparse encoding version [164].

14.3 *HITS*: HUBS AND AUTHORITIES

What is the most important node in a social network? SVD and spectral methods can also help with this question. Next we describe the *HITS* method (also known as 'hubs-and-authorities'), [165], and in the next part, the *PageRank* method, that was the basis behind the extremely successful ranking method of Google [59].

Given a directed graph, like the one in Figure 14.4, we want to find the most important node. The setting was motivated by web search. For example, suppose we search for the keyword "chess," and that the webpages that contain that keyword are the ones of Figure 14.4.

There are two ideas behind *HITS*:

- (a) each node i is assigned two scores, an "authoritativeness" score, a_i, and a "hubness" score h_i

- (b) these scores are defined recursively, in terms of the scores of the neighbors.

The first score, a_i, is the *authoritativeness* score of this node, that is, how good this node is as a source of information for the search keyword. The second score, h_i, is how good this node is as a hub of information.

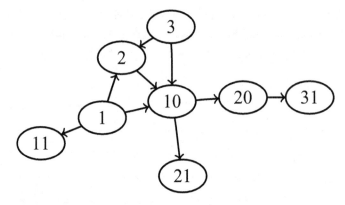

Figure 14.4: Which node is most important/central? *HITS* gives the answer.

The beauty of the approach is that the authoritativeness scores a_i's are constitute the first right-singular vector of the given adjacency matrix $\mathbf{A}$, while the hubness scores h_i's are constitute the first left-singular vector of $\mathbf{A}$.

Observation 14.11   If $\mathbf{A}$ is the adjacency matrix of a set of nodes, then the most central nodes are the ones with the highest authoritativeness scores a_i, where $\vec{a} = (a_1, \ldots, a_N)$ is the right-singular vectors of the SVD decomposition of $\mathbf{A}$.

Justification See [165]. Here is a sketch of the proof: Let $F(i)$ be the set of forward neighbors of i, and $B(i)$ be the set of backward neighbors. Then, *HITS* naturally gives the following recursive definitions:

$$a_i \propto \sum_{j \in B(i)} h_j \tag{14.4}$$

$$h_i \propto \sum_{j \in F(i)} a_i \tag{14.5}$$

That is, node i is a good hub (high h_i) if it points to good authorities; and it is a good authority if it is pointed to by good hubs. Put in matrix and vector form, this says that

$$\vec{h} \propto \mathbf{A}\vec{a} \tag{14.6}$$
$$\vec{a} \propto \mathbf{A}^t\vec{h} \tag{14.7}$$

This is exactly the definition of a pair of left- and right-singular vectors for matrix $\mathbf{A}$. The proportionality constants are related to the singular values of $\mathbf{A}$.

Discussion: It is a pleasant surprise that the circular definition of hubs and authorities eventually has a fixed point. The second pleasant surprise is that we can quickly compute the values a_i, h_i ($i = 1, \ldots, n$), by just iterating equations (Eq. 14.6,14.7) for about ten times (and normalizing to one the length of the vectors after each iteration, to avoid arithmetic overflows). The reason for the convergence is that the first singular value λ_1 of $\mathbf{A}$ is usually much higher than the second, and the ratio determines the convergence speed.

There are several more details on the *HITS* algorithm, but the main point of this section is that SVD can help find important nodes (authorities) and important hubs, in a directed graph.

14.4 PAGERANK

The idea behind PageRank [59] is similar, at the high level: Given a graph of nodes (websites) pointing to each other, a web site is important if important sites point to it. The intuition is nicely described by the *random surfer* setting. Imagine a user moving from site to site, following its links with equal probability, but with occasional restarts. More specifically, if the user is at node i, which has out-degree d_i, then the user either follows one of the d_i neighbors (with probability c), or she "flies out" to a random node (among the N available nodes), with probability $1 - c$. If a node has out-degree d_i =0, then we have immediate teleportation: the surfer flies out randomly to one of the N nodes. The literature refers to this operation as "flying out," or "teleportation."

The *importance* of node i is the steady-state probability p_i, that is, the proportion of time that the surfer will spend on node i.

Arithmetic Example Let's see an example. Consider the network of web sites of Figure 14.5(a), and their to-from column-normalized transition matrix $\mathbf{B}$ in Figure 14.5(b). For the moment, let's assume no teleportation. In that case, the steady state probabilities $\vec{p}$ will satisfy the matrix equation $\mathbf{B}\vec{p} = \vec{p}$. Running the computations, we obtain $\vec{p} = [0.480, \ 0.640, \ 0.480, \ 0.160, \ 0.320]^t$.

Why do we need teleportation, then? The answer is very subtle: the network of Figure 14.5(a) is carefully constructed to form a so-called "ergodic" Markov Chain, which, among other properties, it has no dead ends (that is, no nodes of zero out-degree). In a real setting, there are high chances for dead-end nodes, that would absorb the "random surfer" and keep her there forever. This is exactly what teleportation avoids: when the surfer reaches a dead-end, she teleports randomly to one of the N nodes, with probability $1/N$. The second, also subtle, benefit of teleportation is that it makes the transition matrix *irreducible*: every node can reach every other node, and the Perron-Frobenius theorem applies (Theorem 14.3). Let's see the detailed solution formula.

Mathematical formulas Let $\mathbf{A}$ be the $N \times N$ adjacency matrix, as before. We need a few more definitions:

- $\vec{p} = (p_1, \ldots, p_N)^t$ is the column vector of the steady-state probabilities, that we are looking for

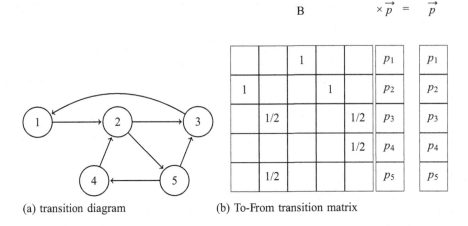

(a) transition diagram   (b) To-From transition matrix

Figure 14.5: Arithmetic example for PageRank: (a) gives the websites and their hypertext links; (b) is the corresponding "To-From" transition matrix. For simplicity, it has no restarts. The 5x1 column vector $\vec{p}$ gives the steady-state probabilities, which correspond to the importance of each node.

- $\mathbf{B}$ is an $N \times N$ matrix, defined as the transposed, column-normalized adjacency matrix: $\mathbf{B} \equiv \mathbf{A}^t * \mathbf{D}^{-1}$ (where $\mathbf{D}$ is a diagonal matrix, with its $d_{i,i}$ element being the degree d_i of node i; if $d_i = 0$, that is, node i is a dead end, then the corresponding column of $\mathbf{B}$ is full of $1/N$ values.)

- $\vec{1}$ is an $N \times 1$ column vector, full of '1's

  Then, the steady-state probability vector $\vec{p}$ is given by

$$\vec{p} = (1 - c)/N[\mathbf{I} - c\mathbf{B}]^{-1} \times \vec{1} \tag{14.8}$$

Connection to eigenvalues   Not surprising, since $\vec{p}$ is the steady state probability vector, it is the eigenvalue of a suitable matrix. Specifically, if we define the modified transition matrix $\mathbf{M}$ as

$$\mathbf{M} = c\mathbf{B} + (1 - c)/N\vec{1} \times \vec{1}^T \tag{14.9}$$

then we have

$$\vec{p} = \mathbf{M} \times \vec{p} \tag{14.10}$$

This is exactly the definition of the first eigenvector of the matrix $\mathbf{M}$, with eigenvalue $\sigma_1 = 1$ (which is expected, since the matrix is column-normalized [253]). The reason that it has a real-valued solution is thanks to the Perron-Frobenius theorem 14.3: the tele-portation is exactly what makes the matrix $\mathbf{M}$ *irreducible*, because it allows every node to connect to every other node.

Arithmetic example – continued   For the network of Figure 14.5(a), and with teleportation probability $1 - c = 0.15$ (which is the value recommended in the literature) the node importance-scores are

$\vec{p} = [0.467, 0.635, 0.475, 0.205, 0.333]^t$ close to the ones without teleportation, and with node "2" being the most central node, again.

***Intuition behind* M**   The intuition behind the transition matrix **M** is that the $m_{i,j}$ element gives the probability that the surfer will transition from node j to node i, when teleportation is possible.

CHAPTER 15

Tensors

15.1 INTRODUCTION

Motivation: Until now we studied graphs that have only one type of edges, and can be represented by an adjacency matrix. What should we do when edges have different types? For example, we can have a graph of who-calls-whom, another graph of who-texts-whom, a third graph of who-works-with-whom. These types of graphs are occasionally referred to as "non-homogeneous" graphs in the literature.

Another example is from network intrusion analysis, where we have triplets of the form

(source-IP, destination-IP, destination-port)

and where the destination port carries important information: for example, port 80 is the http port, giving evidence that the destination machine is probably a web server; high number ports are usually used for peer-to-peer file sharing; an arbitrary port number might indicate that the source machine is doing port-scanning.

How can we find patterns, groups, and anomalies in such a setting? What should we do if we have a fourth piece of information (say, the timestamp, in the computer-network example above)?

15.2 MAIN IDEAS

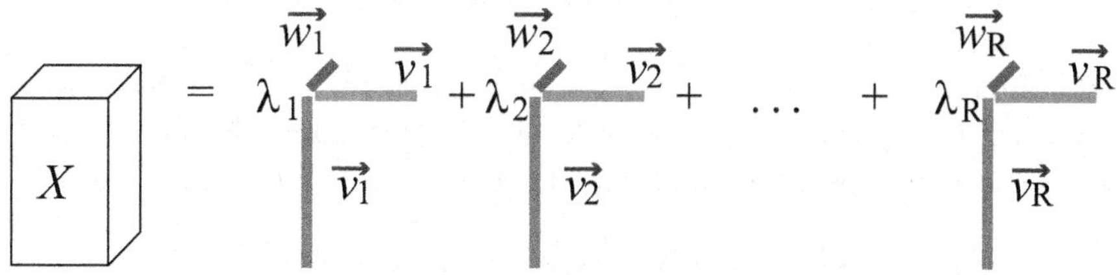

Figure 15.1: Illustration of the PARAFAC (or Kruskal) decomposition: the input three mode tensor on the left is decomposed into R triplets of vectors on the right, reminiscing of the rank-R singular value decomposition of a matrix.

Intuition: This setting is exactly the setting for tensors and tensor decompositions. In short, a tensor is a higher order generalization of a matrix (and vector, and scalar, which are, respectively, tensors of order 2, 1, 0). Figure 15.1 gives an example of a tensor $\mathcal{X}$ – following the convention in the literature, we use calligraphic upper-case font to denote tensors.

The *order* of the tensor is the number of *modes* or *aspects* – in the example of Figure 15.1, the order is 3. The first mode, corresponding to rows, could be "source-IP" in the running example; the second mode (columns) could be "destination-IP" and the third mode (referred to as "tubes") could be the "destination-port."

There are many tensor decompositions [170] but the Kruskal, or PARAFAC, decomposition is the simplest to describe, and it is the one depicted in Figure 15.1. Recall the spectral decomposition version of SVD (Figure 14.3), where we approximate the original matrix with the outer product of one pair, two pairs, etc. of $\vec{u}$ and $\vec{v}$ vectors: what we do here is we approximate the initial tensor $\mathcal{X}$ with the outer product of triplets of vectors $(\vec{u}, \vec{v}, \vec{w})$.

Formulas: We will restrict the discussion on mode-3 tensors, for ease of presentation. However, all the concepts can be generalized to higher-mode tensors.

Mathematically, a (three-mode) tensor $\mathcal{X}$ consists of $x_{i,j,k}$ elements ($i = 1, \ldots, I$, $j = 1, \ldots J$ and $k = 1, \ldots, K$. The λ_i values correspond to the singular values of the SVD, and each triplet of vectors illustrate the soft-membership of the corresponding row/column/tube to the current group ('concept').

To reconstruct the tensor, we use the equation

$$x_{i,j,k} = \sum_{r=1}^{R} \lambda_r * u_{i,r} * v_{j,r} * w_{k,r} \tag{15.1}$$

Discussion: There is a lot of work in tensors, for web analysis [173, 257], information retrieval [79], epilepsy forecasting [1], social network analysis [29], face recognition [274], fMRI analysis [39], chemometrics [60], and more. There are several variations, in addition to the PARAFAC that we describe here (see [174] for a survey), as well as memory-efficient implementations [172] sparse-encoding versions [228].

The interpretation of the triplet vectors is similar to the interpretation of the SVD $\vec{u}_r$ and $\vec{v}_r$ vectors: they provide a form of soft clustering, by indicating how much the respective row/column/tube participates in the corresponding (r-th) concept.

15.3 AN EXAMPLE: TENSORS AT WORK

Consider the setting of [256]: We have sensor data from the Intel Berkeley lab, with I=54 sensors, each capable of measuring K=4 quantities (humidity, temperature, light-intensity, and battery voltage), and we collect measurements for J=1093 time-ticks. Figure 15.3(a) gives the floor-plan, with black circles indicating the sensor positions. Can we find patterns, like major trends, and anomalies?

Tensor analysis can help us do exactly that. Our setting is illustrated in Figure 15.2. The resulting vectors $\vec{u}_i$, $\vec{v}_i$, and $\vec{w}_i$ ($i = 1, 2$, in our case) can help us understand our data, as follows. Notice that we omit some details from [256], for ease of exposition.

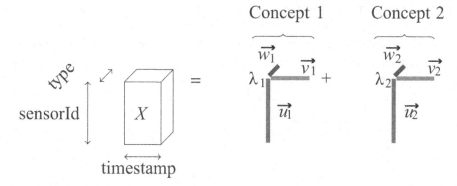

Figure 15.2: Tensor analysis for the SENSOR dataset: Three modes (sensor-id, sensor-type, timestamp), and first two strongest "concepts:" the main trend, and the "close-to-AC" trend (see text).

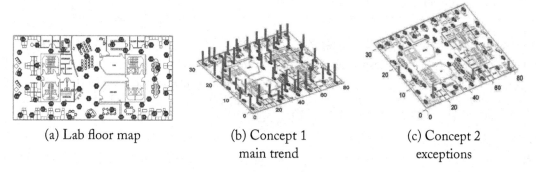

(a) Lab floor map

(b) Concept 1
main trend

(c) Concept 2
exceptions

Figure 15.3: Interpretation of the $\vec{u}$ vectors: (a) shows the floor-plan; (b)/(c) show the score of each sensor, for the first/second "concept" ($\vec{u}$ vectors). Blue/red bars denote positive/negative scores. The first concept is the main, global trend, with vector $\vec{u}_1$ (in (b)): all sensors have high positive scores. Second concept, in (c) negative weights for the bottom left corner and negligible, positive weights to the rest. These few sensors exhibit exceptional behavior.

First mode: $\vec{u}$ ***vectors: spatial trends*** For the first aspect (=mode), that is, the sensor-id, the most dominant trend is the same across all locations/sensors. As shown in Figure 15.3(b), the weights (the blue bars) on all locations have about the same height (entries in the $\vec{u}_1$ vector). The second strongest trend is shown in Figure 15.3(c). The red bars on Figure 15.3(c) indicate negative weights on a few locations close to A/C (mainly at the bottom and left part of the room).

Third mode: $\vec{w}$ ***vectors – sensor-measurement trends*** For sensor-measurement ('type') aspect, the dominant trend is shown as the first concept ($\vec{w}_1$), in Figure 15.4. It indicates that 1) the positive correlation among temperature, light intensity, and voltage level, and 2) negative correlation between humidity and the rest. In retrospect, this makes sense: During the day, temperature and light intensity increase (due to the sun), but humidity drops because the A/C turns on. During the night, temperature and light intensity drop but humidity increases because A/C switches off. The voltage is always positively correlated with the temperature due to the design of MICA2 sensors.

The second row in Figure 15.4 corresponds to abnormal behavior, of the few sensors we saw in Figure 15.3(c). This could be due to low battery for those sensors.

Sensor-Type	voltage	humidity	temperature	light-intensity
concept 1	.16	-.15	.28	.94
concept 2	.6	.79	.12	.01

Figure 15.4: measurement-type correlation: main concept: typical behavior – brightness is correlated with temperature, and anti-correlated with humidity, because the A/C is on during the day (low humidity) and off during the night (high humidity). Second concept: deviating from the usual pattern (possibly due to low battery, for a few of the sensors).

Second mode – $\vec{u}$ vectors – temporal patterns   Figure 15.5(a) shows the $\vec{v}_1$ vector for the first 500 timestamps. It corresponds to the major trend, which is a pattern with daily periodicity, Figure 15.5(b) shows the $\vec{v}_2$ vector, describing the second major trend, for 500 timestamps. Notice that all values are close to zero, except for a small window of time, where there was abnormal activity of the few red-valued sensors of Figure 15.3(c).

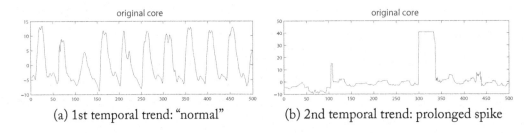

(a) 1st temporal trend: "normal"   (b) 2nd temporal trend: prolonged spike

Figure 15.5: Major temporal trends (third mode, $\vec{w}$ vectors): (a) corresponds to the typical behavior, of daily periodicity, (b) corresponds to $\vec{w}_2$, obviously spotting abnormal behavior: no activity until around time=300, and then a prolonged spike for about 40 time-ticks.

15.4   CONCLUSIONS—PRACTITIONER'S GUIDE

Tensors are powerful tools, capable of doing soft-clustering in all three (or more) modes of the given data cube. Tools to do tensor decomposition are available, some in open-source, like the *Matlab tensor toolbox* [30] with an excellent article by the same authors, Bader and Kolda [28].

This is an active area of research, especially in the effort to make them scalable to large, disk-resident datasets like the GigaTensor algorithm [152], that runs on top of *hadoop*.

CHAPTER 16

Community Detection

Intuitively, a community is a set of nodes where each node has more connections within the community than outside it. This effect has been found (or is believed to exist) in many real-world graphs, especially social networks: Moody [212] finds groupings based on race and age in a network of friendships in one American school;, Schwartz and Wood [244] group people with shared interests from email logs; Borgs et al. [57] find communities from "cross-posts" on Usenet; and Flake et al. [117] discover communities of webpages in the WWW.

We divide the discussion into two parts. First, we will describe the *clustering coefficient*, which is one particular measure of community structure that has been widely used in the literature. Next, we will describe methods for *extracting* community structure from large graphs.

16.1 CLUSTERING COEFFICIENT

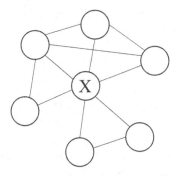

Figure 16.1: *Clustering coefficient:* Node X has $k_X = 6$ neighbors. There are only $n_X = 5$ edges between the neighbors. So, the local clustering coefficient of node X is $n_X/k_X = 5/15 = 1/3$.

Informal description: The clustering coefficient measures the "clumpiness" of a graph, and has relatively high values in many graphs.

Detailed description: We will first define the clustering coefficient for one node, following [279] and [214]:

Definition 16.1 Clustering Coefficient   Suppose a node i has k_i neighbors, and there are n_i edges between the neighbors. Then the clustering coefficient of node i is defined as

$$C_i = \begin{cases} \frac{n_i}{k_i} & k_i > 1 \\ 0 & k_i = 0 \text{ or } 1 \end{cases} \qquad (16.1)$$

Thus, it measures the degree of "transitivity" of a graph; higher values imply that "friends of friends" are themselves likely to be friends, leading to a "clumpy" structure of the graph. See Figure 16.1 for an example.

For the clustering coefficient of a *graph* (the *global* clustering coefficient), there are two definitions:

1. Transitivity occurs iff *triangles* exist in the graph. This can be used to measure the global clustering coefficient as

$$C = \frac{3 \times \text{number of triangles in the graph}}{\text{number of connected triples in the graph}} \qquad (16.2)$$

   where a "connected triple" is a triple of nodes consisting of a central node connected to the other two; the flanking nodes are unordered. The equation counts the fraction of connected triples which actually form triangles; the factor of three is due to the fact that each triangle corresponds to three triples.

2. Alternatively, Watts and Strogatz [279] use equation 16.1 to define to a *global* clustering coefficient for the graph as

$$C = \sum_{i=1}^{N} C_i / N \qquad (16.3)$$

The second definition leads to very high variance in the clustering coefficients of low-degree nodes (for example, a degree 2 node can only have $C_i = 0$ or 1). The results given by the definitions can actually be quite different. The first definition is usually easier to handle analytically, while the second one has been used extensively in numerical studies.

Computation of the clustering coefficient: Alon et al. [22] describe a deterministic algorithm for counting the number of triangles in a graph. Their method takes $O(N^\omega)$ time, where $\omega < 2.376$ is the exponent of matrix multiplication (that is, matrix multiplication takes $O(N^\omega)$ time). However, this is more than quadratic in the number of nodes, and might be too slow for large graphs. Bar-Yossef et al. [34] describe algorithms to count triangles when the graph is in *streaming* format, that is, the data is a stream of edges which can be read only sequentially and only once. The advantage of streaming algorithms is that they require only one pass over the data, and so are very fast; however, they typically require some temporary storage, and the aim of such algorithms is to minimize this space requirement. They find an $O(\log N)$-space randomized algorithm for the case when the edges are sorted on the source node. They also show that if there is no ordering on the edges, it is impossible to count the number of triangles using $o(N^2)$ space.

Clustering coefficients in the real world: The interesting fact about the clustering coefficient is that it is almost always larger in real-world graphs than in a random graph with the same number of nodes and edges (random graphs are discussed later; basically these are graphs where there are no biases toward any nodes). Watts and Strogatz [279] find a clustering coefficient of 0.79 for the actor network (two actors are linked if they have acted in the same movie) whereas the corresponding random graph has a coefficient of 0.00027. Similarly, for the power grid network, the coefficient is 0.08, much greater than 0.005 for the random graph.

Extension of the clustering coefficient idea: While the global clustering coefficient gives an indication of the overall "clumpiness" of the graph, it is still just one number describing the entire graph. We can look at the clustering coefficients at a finer level of granularity by finding the average clustering coefficient $C(k)$ for all nodes with a particular degree k. Dorogovtsev et al. [96] find that for scale-free graphs generated in a particular fashion, $C(k) \propto k^{-1}$. Ravasz and Barabási [236] investigate the plot of $C(k)$ versus k for several real-world graphs. They find that $C(k) \propto k^{-1}$ gives decent fits to the actor network, the WWW, the Internet AS level graph and others. However, for certain graphs like the Internet Router level graph and the power grid graph, $C(k)$ is independent of k. The authors propose an explanation for this phenomenon: they say that the $C(k) \propto k^{-1}$ scaling property reflects the presence of hierarchies in the graph. Both the Router and power-grid graphs have geographical constraints (it is uneconomic to lay long wires), and this presumably prevents them from having a hierarchical topology.

16.2 METHODS FOR EXTRACTING GRAPH COMMUNITIES

The problem of extracting communities from a graph, or of dividing the nodes of a graph into distinct communities, has been approached from several different directions. In fact, algorithms for "community extraction" have appeared in practically all fields: social network analysis, physics, and computer science among others. Here, we collate this information and present the basic ideas behind some of them. The computational requirements for each method are discussed alongside the description of each method. A survey specifically looking at clustering problems from bioinformatics is provided in [196], though it focuses only on bipartite graphs.

Dendrograms: Traditionally, the sociological literature has focused on communities formed through *hierarchical clustering* [109]: nodes are grouped into hierarchies, which themselves get grouped into high-level hierarchies and so. The general approach is to first assign a value V_{ij} for every pair (i, j) of nodes in the graph. Note that this value is different from the *weight* of an edge; the weight is a part of the data in a weighted graph, while the value is computed based on some property of the nodes and the graph. This property could be the distance between the nodes, or the number of node-independent paths between the two nodes (two paths are node-independent if the only nodes they share are the endpoints). Then, starting off with only the nodes in the graph (with no edges included), we add edges one by one in decreasing order of value. At any stage of this algorithm, each of the connected components corresponds to a community. Thus, each iteration of this algorithm represents a set of communities; the *dendrogram* is a tree-like structure, with the individual nodes of the graph as the leaves of the tree and the communities in each successive iteration being the internal nodes of the tree. The root node of the tree is the entire graph (with all edges included).

While such algorithms have been successful in some cases, they tend to separate fringe nodes from their "proper" communities [127]. Such methods are also typically costly; however, a carefully constructed variation [81] requires only $O(Ed \log N)$ time, where E is the number of edges, N the number of nodes, and d the depth of the dendrogram.

Edge-betweenness or Stress: Dendrograms build up communities from the bottom up, starting from small communities of one node each and growing them in each iteration. As against this, Girvan and Newman [127] take the entire graph and remove edges in each iteration; the connected components in each stage are the communities. The question is: how do we choose the edges to remove? The authors remove nodes in decreasing order of their "edge-betweenness," as defined below.

Definition 16.2   Edge-betweenness or Stress   Consider all shortest paths between all pairs of nodes in a graph. The edge-betweenness or stress of an edge is the number of these shortest paths that the edge belongs to.

The idea is that edges connecting communities should have high edge-betweenness values because they should lie on the shortest paths connecting nodes from different communities. Tyler et al. [271] have used this algorithm to find communities in graphs representing email communication between individuals.

The edge-betweenness of all edges can be calculated by using breadth-first search in $O(NE)$ time; we must do this once for each of the E iterations, giving a total of (NE^2) time. This makes it impractical for large graphs.

Goh et al. [129] measure the distribution of edge-betweenness, that is, the count of edges with an edge-betweenness value of v, versus v. They find a power law in this, with an exponent of 2.2 for protein interaction networks, and 2.0 for the Internet and the WWW.

Max-flow min-cut formulation: Flake et al. [117] define a community to be a set of nodes with more intra-community edges than inter-community edges. They formulate the community-extraction problem as a minimum cut problem in the graph; starting from some *seed* nodes which are known to belong to a community, they find the minimal-cut set of edges that disconnects the graph so that all the seed nodes fall in one connected component. This component is then used to find new seed nodes; the process is repeated till a good component is found. This component is the community corresponding to the seed nodes.

One question is the choice of the original seed nodes. The authors use the HITS algorithm [165], and choose the hub and authority nodes as seeds to bootstrap their algorithm. Finding these seed nodes requires finding the first eigenvectors of the adjacency matrix of the graph, and there are well-known iterative methods to approximate these [45]. The min-cut problem takes polynomial time using the Ford-Fulkerson algorithm [87]. Thus, the algorithm is relatively fast, and is quite successful in finding communities for several datasets.

Graph partitioning: A very popular clustering technique involves graph partitioning: the graph is broken into two partitions or communities, which may then be partitioned themselves. Several different measures can be optimized for while partitioning a graph. The popular METIS software tries to find the best separator, minimizing the number of edges cut in order to form two disconnected components of relatively similar sizes [157]. Other common measures include *coverage* (ratio of intra-cluster edges to the total number of edges) and *conductance* (ratio of inter-cluster edges to a weighted function of partition sizes) [58]. Detailed discussions on these are beyond the scope of this work.

Several heuristics have been proposed to find good separators; *spectral clustering* is one such highly successful heuristic. This uses the first few singular vectors of the adjacency matrix or its *Laplacian* to partition the graph (the Laplacian matrix of an undirected graph is obtained by subtracting its adjacency matrix from a diagonal matrix of its vertex degrees) [21, 252]. Kannan et al. [156] find that spectral heuristics give good separators in terms of both coverage and conductance. Another heuristic method called *Markov Clustering* [273] uses random walks, the intuition being that a random walk on a dense cluster will probably not leave the cluster without visiting most of its vertices. Brandes et al. [58] combine spectral techniques and minimum spanning trees in their GMC algorithm.

In general, graph partitioning algorithms are slow; for example, spectral methods taking polynomial time might still be too slow for problems on large graphs [156]. However, Drineas et al. [101] propose combining spectral heuristics with fast randomized techniques for singular value decomposition to combat this. Also, the *number* of communities (e.g., the number of eigenvectors to be considered in spectral clustering) often needs to be set by the user, though some methods try to find this automatically [43, 263].

Bipartite cores: Another definition of "community" uses the concept of *hubs* and *authorities*. According to Kleinberg [165], each hub node points to several authority nodes, and each authority node is pointed to by several hub nodes. Kleinberg proposes the HITS algorithm to find such hub and authority nodes. Gibson et al. [125] use this to find communities in the WWW in following fashion. Given a user query, they use the top (say, 200) results on that query from some search engine as the seed nodes. Then they find all nodes linking to or linked from these seed nodes; this gives a subgraph of the WWW which is relevant to the user query. The HITS algorithm is applied to this subgraph, and the top ten hub and authority nodes are together returned as the core community corresponding to the user query.

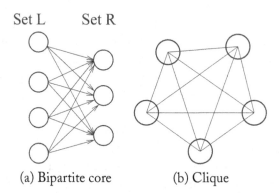

Set L   Set R

(a) Bipartite core   (b) Clique

Figure 16.2: *Indicators of community structure:* Plot (a) shows a 4 × 3 bipartite core, with each node in Set *L* being connected to each node in Set *R*. Plot (b) shows a five-node clique, where each node is connected to every other node in the clique.

Kumar et al. [179] remove the requirement for a user query; they use *bipartite cores* as the seed nodes for finding communities. A bipartite core in a graph consists of two (not necessarily disjoint) sets of nodes *L* and *R* such that every node in *L* links to every node in *R*; links from *R* to *L* are optional (Figure 16.2). They describe an algorithm that uses successive elimination and generation phases to generate bipartite cores of larger sizes each iteration. As in [125], these cores are extended to form communities using the HITS algorithm.

HITS requires finding the largest eigenvectors of the $A^t A$ matrix, where A is the adjacency matrix of the graph. This is a well-studied problem. The elimination and generation passes have bad worst case complexity bounds, but Kumar et al. [179] find that it is fast in practice. They attribute this to the strongly skewed distributions in naturally occurring graphs. However, such algorithms which use hubs and authorities might have trouble finding *webrings*, where there are no clearly defined nodes of "high importance."

Local Methods: All the previous techniques used *global* information to determine clusters. This leads to scalability problems for many algorithms. Virtanen [275] devised a clustering algorithm based solely on *local* information derived from members of a cluster. Defining a fitness metric for any cluster candidate, the author uses simulated annealing to locally find clusters which approximately maximize fitness. The advantage of this method is the online computation of locally optimal clusters (with high probability) leading to good scalability, and the absence of any "magic" parameters in the algorithm. However, the memory and disk access requirements of this method are unclear.

Cross-Associations: Recently, Chakrabarti et al. [71] (also see [69]) devised a scalable, parameter-free method for clustering the nodes in a graph into groups. Following the overall MDL (Minimum Description Length) principle, they define the goodness of a clustering in terms of the quality of lossless compression that can be attained using that clustering. Heuristic algorithms are used to find good clusters of nodes, and also to automatically determine the *number* of node clusters. The algorithm is linear in the number of edges E in the graph, and is thus scalable to large graphs.

Communities via Kirchoff's Laws: Wu and Huberman [284] find the community around a given node by considering the graph as an electric circuit, with each edge having the same resistance. Now, one Volt is applied to the given node, and zero Volts to some other randomly chosen node (which will hopefully be outside the community). The voltages at all nodes are then calculated using Kirchoff's Laws, and the nodes are split into two groups by (for example) sorting all the voltages, picking the median voltage, and splitting the nodes on either side of this median into two communities. The important idea is that the voltages can be calculated approximately using iterative methods requiring only $O(N + E)$ time, but with the quality of approximation depending *only* on the number of iterations and not on the graph size.

This is a fast method, but picking the correct nodes to apply zero Volts to is a problem. The authors propose using randomized trials with repetitions, but further work is needed to prove formal results on the quality of the output.

16.3 A WORD OF CAUTION—"NO GOOD CUTS"

Despite the wealth of graph partitioning algorithms, the reader should be aware that, often, real graphs have *no good cuts*. For example, consider a full clique: there is no good way to cut it in half. Similarly, neither a star with n satellites, nor a *jelly fish*-type graph (from Section 5.1, p. 31), as shown in Figure 5.1, have a good cut.

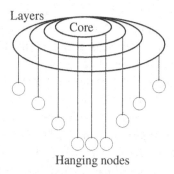

Figure 16.3: *The Internet as a "Jellyfish:"* The Internet AS-level graph can be thought of as a core, surrounded by concentric layers around the core. There are many one-degree nodes that hang off the core and each of the layers.

This observation has been made several times, independently [68], [70], [187]. One of the most striking observations is the so-called "min-cut" plot, shown in Figure 16.4. This plots the normalized size of the cut E_{cut} (fraction of edges deleted), versus the size of the graph (count of nodes N). For a 2 d grid, the relationship is $E_{cut} \propto N^{1-1/2}$. In general, for a D-dimensional grid, the equation is

$$E_{cut}/N \propto N^{-1/D}$$

which means the slope of E_{cut}/N versus N on the log-log scale is always negative, and it only approaches zero for, say, a large, random graph (which approaches dimensionality of infinity).

What happens on real graphs is astonishing: the dimensionality is *negative*, as indicated by the positive slope at the rightmost part of the plots, for several real graphs. This means that our intuition about cuts and communities is completely wrong, if we are based on low-dimensionality grids.

From a practical point of view, this means that for real graphs, we might not be able to find good global cuts, no matter how good is the graph-partitioning algorithm we use. Next we elaborate on this subtle, but vital observation.

"Min-cut plots:"   In more detail, a min-cut of a graph $\mathcal{G} = (\mathcal{V}, \mathcal{E})$ is a partition of the set of vertices $\mathcal{V}$ into two sets $\mathcal{V}_1$ and $\mathcal{V} - \mathcal{V}_1$ such that both partitions are of approximately the same size, and the number of edges crossing partition boundaries is minimized. The number of such edges in the min-cut is called the *min-cut size*. Min-cut sizes of various classes of graphs has been studied extensively, because they influence other properties of the graphs [241]. For example, Figure 16.4(a) shows a regular 2D grid graph, and one possible min-cut of the graph. Notice that if the number of nodes is N, then the size of the min-cut (in this case) is $O(\sqrt{N})$.

The min-cut plot is built as follows: given a graph, its min-cut is found, and the set of edges crossing partition boundaries deleted. This divides the graph into two disjoint graphs; the min-cut algorithm is then applied recursively to each of these sub-graphs. This continues till the size of the graph reaches a small value (set to 20 in our case). Each application of the min-cut algorithm becomes a point in the min-cut plot. The graphs are drawn on a log-log scale. The x-axis is E, the number of edges in a given graph. The y-axis is E_{cut}/E, the fraction of that graph's edges that were included in the edge-cut for that graph's separator.

Figure 16.4(b) shows the min-cut plot for the 2D grid graph. In plot (c), the value on the y-axis is averaged over all points having the same x-coordinate. The min-cut size is $O(\sqrt{N})$, so this plot should have a slope of -0.5, which is exactly what we observe.

Figure 16.5 shows min-cut sizes of some real-world graphs. For random graphs, we expect about half the edges to be included in the cut. Hence, the min-cut plot of a random graph would be a straight horizontal line with a y-coordinate of about $\log(0.5) = -1$. A very separable graph (for example, a line graph) might have only one edge in the cut; such a graph with N edges would have a y-coordinate of $\log(1/N) = -\log(N)$, and its min-cut plot would thus be on the line $y = -x$. As we can see from Figure 16.5, the plots for real-world graphs do not match either of these situations, meaning that real-world graphs are quite far from either random graphs or simple line graphs.

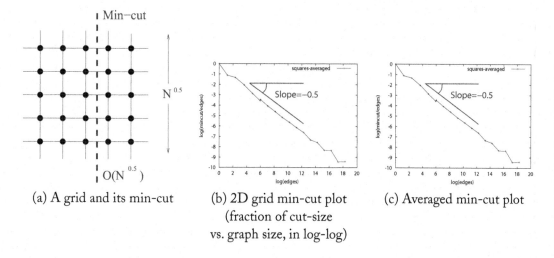

(a) A grid and its min-cut

(b) 2D grid min-cut plot
(fraction of cut-size
vs. graph size, in log-log)

(c) Averaged min-cut plot

Figure 16.4: Plot (a) shows a portion of a regular 400x400 2D grid, and a possible min-cut. Plot (b) shows the full min-cut plot, and plot (c) shows the averaged plot. If the number of nodes is N, the length of each side is $\sqrt{N}$. Then the size of the min-cut is $O(\sqrt{N})$; this leads to a slope of -0.5, which is exactly what we observe.

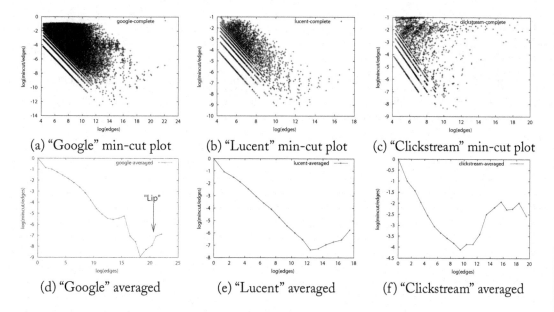

(a) "Google" min-cut plot

(b) "Lucent" min-cut plot

(c) "Clickstream" min-cut plot

(d) "Google" averaged

(e) "Lucent" averaged

(f) "Clickstream" averaged

Figure 16.5: These are the min-cut plots for several datasets. We plot the ratio of mincut-size to edges versus number of edges on a log-log scale. The first row shows the actual plots; in the second row, the cutsize-to-edge ratio is averaged over all points with the same number of edges.

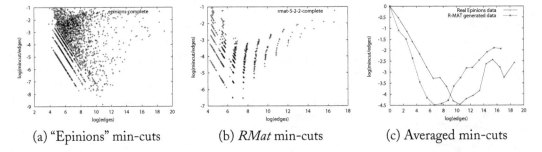

(a) "Epinions" min-cuts   (b) *RMat* min-cuts   (c) Averaged min-cuts

Figure 16.6: Here, we compare min-cut plots for the Epinions dataset and a dataset generated by *RMat*, using properly chosen parameters (in this case, a=0.5, b=0.2, c=0.2, d=0.1). We see from plot (c) that the shapes of the min-cut plots are similar.

CHAPTER 17

Influence/Virus Propagation and Immunization

For a given, arbitrary graph, what is the epidemic threshold? That is, what are the conditions under which a virus (or rumor, or new product) will result in an epidemic, taking over the majority of the population as opposed to dying out? Prakash et al. [232] give the *super-model* theorem which generalizes older results in two important, orthogonal dimensions. The theorem shows that

- (a) for a wide range of *virus propagation models* (VPM) that include *all* virus propagation models in standard literature (say, [140][105]), and
- (b) for *any* contact graph, the answer always depends on the first eigenvalue of the connectivity matrix.

Here we give the theorem, arithmetic examples for popular VPMs (=Virus Propagation Models), like flu (SIS), mumps (SIR), SIRS, and more. We also discuss the implications of the theorem: easy (although sometimes *counter-intuitive*) answers to "what-if" questions, easier design and evaluation of immunization policies, and significantly faster agent-based simulations.

17.1 INTRODUCTION—TERMINOLOGY

Given a social or computer network, where the links represent who has the potential to infect whom, can we tell whether a virus will create an epidemic, as opposed to quickly becoming extinct? This is a fundamental question in epidemiology; intuitively, the answer should depend on (a) the graph connectivity, and (b) the virus propagation model (VPM), that is, how virulent is it, how quickly the host recovers (if ever), whether the host obtains (or is born with) immunity, how quickly (if ever) the host loses immunity, etc. This threshold is the level of virulence below which a virus is guaranteed of dying out quickly [168].

The overwhelming majority of earlier work focuses either on full-clique topologies (everybody contacts everybody else), or on "homogeneous" graphs [160, 161], or on power-law graphs [229] or hierarchical (near-block-diagonal) topologies [141] (people within a community contact all others in this community, with a few cross-community contacts). The only exception that examine arbitrary-topology graphs is [276] and its follow-up work [67], [120] which all focused on only a single model, the "SIS" one (flu-like, with no immunity).

The upcoming result shows that, irrespective of the virus propagation model, the effect of the underlying topology can be captured by just one parameter: the first eigenvalue λ_1 of the adjacency matrix **A**. In particular, it covers *all* models given in the standard survey by Hethcote [140], which includes models like SIS (no immunity, like flu – "susceptible, infected, susceptible") and SIR (life-time immunity, like mumps: "susceptible, infected, recovered"). It also includes numerous other cases like SIRS [105] (temporary immunity), as well as some useful generalizations SIV (vigilance/vaccination with temporary immunity) and SEIV (vigilance/vaccination with temporary immunity *and* virus incubation) and many more. A few of these models are shown in Figure 17.4, organized in a lattice: Models in child nodes are special cases of the model in the parent node.

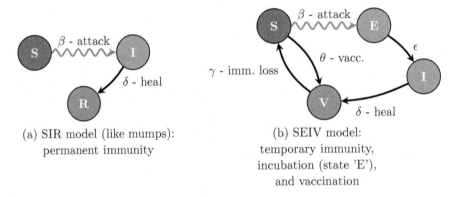

(a) SIR model (like mumps):
permanent immunity

(b) SEIV model:
temporary immunity,
incubation (state 'E'),
and vaccination

Figure 17.1: Examples of virus propagation models: (a) SIR, like mumps, with life-time immunity. (b) SEIV – more complicated: temporary immunity (lost with prob. γ), vaccination (with prob. θ), and "Exposed" (but not infectious – yet) state. Both models and many more, follow the first-eigenvalue threshold.

Informally, the main result is:

Informal Theorem 1 *For any virus propagation model (VPM) in the published literature, operating on an underlying undirected contact-network of any arbitrary topology with adjacency matrix* **A**, *the epidemic threshold depends only on the first eigenvalue*

$$\lambda_1$$

of **A** *and some constant* C_{VPM} *that is determined by the virus propagation model.*

In a nutshell, the typical states are the following, out of which the first three are the most typical ones:

- 'S:' Susceptible/healthy
- 'I:' Infected (and infectious)
- 'R:' Removed/recovered – the node has immunity for life (or is deceased)
- 'V:' Vigilant: the node can not be infected (but may lose it's immunity, depending on the VPM)
- 'E:' Exposed: the node is not infectious, but it is a carrier of the virus, and it will eventually evolve to the "Infected/Infectious" state.

- 'M:' mother-inherited immunity, like a newborn that initially carries the mother's antibodies, but the node will eventually evolve to the "susceptible" state.

Also notice that we have a whole class of infected states (like 'I,' 'E'), where the node has the virus; a whole class of vigilant states (like 'R,' 'V,' 'M'), where the node is healthy and invincible. The up coming model is very general, exactly allowing full classes of susceptible, infected, and vigilant states, as we show in Figure 17.5.

VPM	virus-propagation model
β	attack/transmission probability over a contact-link
δ	healing probability once infected
γ	immunization-loss probability once recovered (in SIRS) or vigilant (in SIV, SEIV)
ϵ	virus-maturation probability once exposed – hence, $1 - \epsilon$ is the virus-incubation probability
θ	direct-immunization probability when susceptible
$\mathbf{A}$	adjacency matrix of the underlying undirected contact-network
N	number of nodes in the network
λ_1	largest (in magnitude) eigenvalue of $\mathbf{A}$
s	effective strength of a epidemic model on a graph with adjacency matrix $\mathbf{A}$

Figure 17.2: Symbols and definitions

SIS	"susceptible, infected, susceptible" VPM – no immunity, like flu
SIR	"susceptible, infected, recovered" VPM – life-time immunity, like mumps
SIRS	VPM with temporary immunity
SIV	"susceptible, infected, vigilant" VPM – immunization/vigilance with temporary immunity
SEIR	"susceptible, exposed, infected, recovered" VPM – life-time immunity *and* virus incubation
SEIV	VPM with vigilance/immunization with temporary immunity *and* virus incubation

Figure 17.3: Some Virus Propagation Models (VPMs)

Figure 17.4 shows the generalization hierarchy for some common epidemic models. The brown colored nodes denote standard VPMs found in literature while the blue colored nodes denote generalizations introduce in [232]. Each VPM is a generalization of all the models below it, e.g., SIV is a generalization of SIRS, SIR, and SIS. The main generalization, $S^*I^2V^*$, is illustrated in Figure 17.5 and further discussed in [232].

17.2 MAIN RESULT AND ITS GENERALITY

The tipping point for each of the models captures a fundamental transition in the behavior of the system between the two extremes: a network-wide epidemic, versus a minor local disturbance that

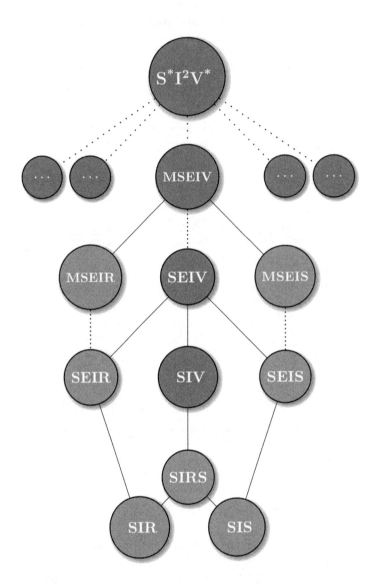

Figure 17.4: Lattice of Virus Propagation models, including SIS (flu); SIR (mumps); SIRS (temporary immunity); SIV (vigilance, i.e., pro-active vaccination); SEIV (like SIV, with virus incubation, i.e., the "exposed but not infectious" state); MSEIR (with the passive immune state M, like newborns that inherit mother's immunity); and the main generalization $S^*I^2V^*$. Each VPM is a generalization of all the models below it.

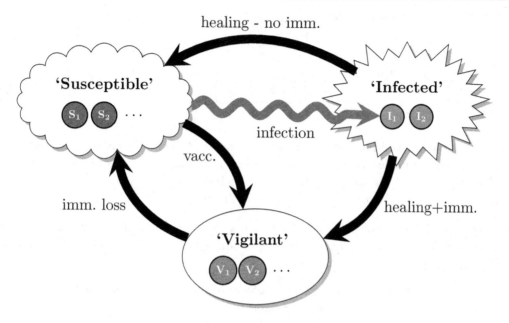

Figure 17.5: Pseudo-State-Diagram for a node in the graph in the generalized model $S^*I^2V^*$. Three types of states – Susceptible (healthy but can get infected), Infected (capable of transmission), and Vigilant (healthy and can't get infected). Within-class transitions not shown for clarity. Red-curvy arrow indicates *graph-based* transition, due to neighbors of the node; all other transitions are *endogenous* (caused by the node itself with some probability at every time step).

fizzles out quickly. We use the typical definition of the threshold used in literature [40, 67, 120, 140]. Intuitively, below threshold the virus has no chance of spreading the infection while, above threshold, the virus may take over and create an epidemic. For the SIS model (=no immunity), the tipping point describes some maximum *strength* of a virus, that will guarantee no epidemic [67, 140]. We define *strength* later (see Equation 17.2). Similarly, for the SIR model, the tipping point relates the explosiveness of the infection phase with respect to the virus strength, since in this model the virus will become extinct.

To standardize the discussion of threshold results, we cast the threshold problem as expressing the normalized *effective strength* of a virus as a function of the *particular* propagation model and the *particular* underlying contact-network. So we are "above threshold" when the effective strength $s > 1$, "under threshold" when $s < 1$ and the threshold or the tipping point is reached when $s = 1$.

Formally, the main result is:

Theorem 17.1   Super-model theorem – sufficient condition for stability   *For virus propagation models that follow the $S^*I^2V^*$ model (see [232]) and for any arbitrary undirected graph with adjacency* *matrix* **A** *the sufficient condition for stability is given by:*

$$s < 1 \tag{17.1}$$

where, s (the effective strength) is:

$$s = \lambda_1 \cdot C_{\text{VPM}} \tag{17.2}$$

with λ_1 being the largest eigenvalue of the adjacency matrix, C_{VPM} is a constant that depends on the virus propagation model (see Figure 17.6). Hence, the tipping point is reached when $s = 1$.

The result generalizes along two different, difficult directions: (a) arbitrary contact-network topologies, and (b) several virus propagation models (VPMs).

General Topologies   Much of previous work [25] has concentrated on the analysis of VPMs on *specific* types of contact-networks, typically cliques or homogeneous graphs. We include them all, as *special* cases. Specifically

- Cliques, where every node contacts every other node. In that case, our result gives $\lambda_1 = N$, where N is the number of nodes in the graph.
- Homogeneous graphs, with fixed degree d and random Erdös-Rényi graphs with expected degree d (e.g., see [160, 161]). In all these cases we have $\lambda_1 = d$, and our theorem includes the previous results.
- Hierarchical (i.e., near-block-diagonal), e.g. [141].
- Power-law random graphs (e.g. [229]).

Theorem 17.1 provides a simple and natural generalization of these results to arbitrary graphs. For example, previous results [229] have shown that the epidemic threshold in case of scale-free (power-law) networks is vanishingly small as the size N of the network increases. This is a corollary of our theorem: When a power-law graph grows ($N \to \infty$), the largest eigenvalue grows with the highest degree, which also grows infinity, and thus the threshold approaches zero.

General VPMs   The result generalizes the threshold results for *any* VPM that is a special case of the $S^* I^2 V^*$ model. We refer to this generalized model as $S^* I^2 V^*$, because it has an arbitrary number of susceptible states, two infectious/infected states, and an arbitrary number of vigilant/vaccinated (= recovered) states. All the standard models (like see [140], [105]) are simply *special* cases of $S^* I^2 V^*$:

- the typical flu model, SIS, is a special case;
- the typical mumps model, SIR, which corresponds to permanent immunity, is a special case, with one state for each class − S belongs to the Susceptible class, I belongs to the Infected class, and R belongs to the Vigilant class:
- the SIRS model (temporary immunity), similar to the SIR model;
- the SEIRS model ([140], page 601) where the virus has an incubation period (state 'E:' exposed, but not infectious), and all other ingredients of the SIRS model (temporary immunity).

We now give a brief summary of the threshold results (Figure 17.6) by applying Theorem 17.1 on some standard epidemic models. Note the effect of the contact-network in effective strength for *each* model is captured solely by one parameter, λ_1, the first eigenvalue of the adjacency matrix of the network. Again, our result is a general one and these models just highlight the ready applicability of the result to standard VPMs in use.

Model	Effective Strength (s)	Threshold
SIS (e.g., flu)	$\lambda_1 \cdot \left(\frac{\beta}{\delta}\right)$	
SIR (e.g., mumps)	$\lambda_1 \cdot \left(\frac{\beta}{\delta}\right)$	
SIRS (e.g., pertussis)	$\lambda_1 \cdot \left(\frac{\beta}{\delta}\right)$	
SIV	$\lambda_1 \cdot \left(\frac{\beta\gamma}{\delta(\gamma+\theta)}\right)$	$s = 1$
SEIR	$\lambda_1 \cdot \left(\frac{\beta}{\delta}\right)$	
SEIV	$\lambda_1 \cdot \left(\frac{\beta\gamma}{\delta(\gamma+\theta)}\right)$	
$SI_1I_2V_1V_2$ (*used to model the H.I.V. virus, e.g., see [25]*)	$\lambda_1 \cdot \left(\frac{\beta_1 v_2 + \beta_2 \epsilon}{v_2(\epsilon + v_1)}\right)$	

Figure 17.6: Threshold results for some models. β, δ, γ are the probabilities to attack, heal, and lose-immunity.

17.3 APPLICATIONS

The results in this chapter can be fundamental to numerous applications. We describe a few important ones next.

Fast answers to "what-if" questions and guiding policy The threshold results can help quickly determine the result of plausible situations. For example, what happens if the virus is twice as infectious (virulent)? Similarly, what happens when there is a weaker strain of the virus? Our results will help in determining whether there is a danger of the infection taking off or not. Naturally then this can feed into policy decisions for controlling epidemics. Assuming some models for the underlying contact network (like scale-free, small-world, hierarchical, etc.) we can estimate which nodes/classes should be quarantined or immunized first. Given the linear dependence on λ_1, we want to immunize nodes (and hence remove them from the contact graph) which will drop the λ_1 value the most so that the resultant infection becomes below threshold and dies out. For example, they may decide to immunize teachers and kindergarten children first to control the epidemic. In addition, they can impose restrictions on travel so as to not increase the λ_1 and hence the effective strength for the virus. The above discussion also illustrates the generality of our result. Policy makers can assume *any* graph model which captures the contact behavior of the population the best and still use our threshold result to guide policy.

A lot of work has been done to show that immunizing high-degree nodes in scale-free networks is a good idea because of the vanishing threshold result [229]. But significantly, just concentrating on high-degree nodes will *miss* those low-degree nodes which are good "bridges" and hence can have an important influence on decreasing λ_1 when immunized. Intuitively, how *disparate* the groups are to which node connects is also important in addition to how *many* groups one is connected to. For example, a single common friend of only some sportspeople and movie stars can have a huge impact on the outbreak of a disease even if he/she knows only a few sportspeople and movie stars (while sportspeople and movie stars are themselves very tightly connected).

We have been concentrating on biological viruses only, but various biological virus models have been used to model computer viruses as well [168], e.g., [138] introduced the SHIR model ("susceptible," "hidden," "infected," "recoverable") to model computers under email attack. In these cases, more so than the biological ones, it is easier to get the entire underlying network. Hence, our threshold results can be precisely used to make the network more robust to malware and computer viruses. This can be done by selectively "removing" nodes from the contact-network by immunizing them like installing a firewall on them, etc.

Immunization policies Another related topic has been on finding the right immunization policy. Pastor-Satorras and Vespignani [230] find that randomly selecting nodes for immunization performs much worse than "targeted" immunization, which selects the nodes with the highest connectivity. This is as expected; removing the highest-degree nodes quickly disconnects the graph [20, 50, 220], preventing the spread of infection.

Theorem 17.1 dictates an optimal immunization policy, namely, try to minimize the eigenvalue of the matrix after immunization. If eigen-drop is the difference in eigenvalue before and after immunization, our goal is to maximize the eigen-drop, *no matter* what is the behavior of the virus is (mumps-like, flu-like, HIV-like, etc.). And, as intuitively expected, immunizing nodes with a high degree usually (but not always) lead to best eigen-drop. A systematic way of choosing nodes to immunize is given in [264] and [267]: The optimal solution is infeasible, and they propose carefully designed heuristics to maximize the eigen-drop, given a fixed budget k of vaccines.

Time-varying graphs   An orthogonal extension is the case that the contact/connectivity matrix change over time, say, with a daily periodicity: During the day, we come in contact with our colleagues, teachers, school-mates; during the night we are only in contact with family members; and similarly, we have different contacts during the weekend (friends, extended family, etc.). It turns out that, at least for the SIS model, we can compute the epidemic threshold [233], and it is again related to the eigenvalue of a product of matrices.

The problem of time-varying graphs is especially of interest for the propagation of computer viruses on mobile phones. The eigenvalue arguments provide solutions to such cases, too [272].

17.4 DISCUSSION

We discuss some simulation examples for the threshold result in some models and a few direct implications of the super-model theorem in this section. We also illustrate what the result implies for the "vulnerability" of the underlying contact graph for epidemics. Apart from the dependence of the threshold on λ_1, it is instructive to note some unexpected results in specific models as well.

17.4.1 SIMULATION EXAMPLES

The work in [232] reports computer simulation experiments on the *Oregon* dataset (see Figure 1.2, p. 4): This is the so-called Oregon autonomous system (AS) router graph, a real network collected

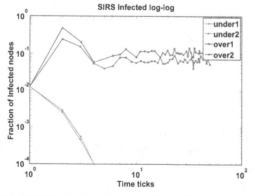
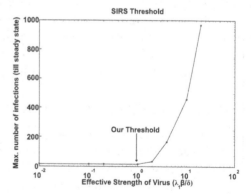
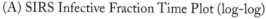

(A) SIRS Infective Fraction Time Plot (log-log)

(B) SIRS Max. Infections till steady state vs Strength (lin-log)

Figure 17.7: SIRS (all values averages over several runs): (A) Plot of Infective Fraction of Population vs Time (log-log). Note the qualitative difference in behavior *under* (green) and *above* (red) the threshold. (B) Plot of Max. number of infected nodes till steady state vs Effective Strength (lin-log). Note the tipping point is exactly when the effective strength $s = 1$.

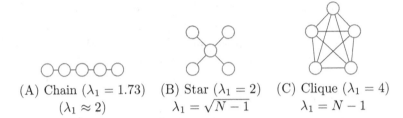

(A) Chain ($\lambda_1 = 1.73$)   (B) Star ($\lambda_1 = 2$)   (C) Clique ($\lambda_1 = 4$)
($\lambda_1 \approx 2$) $\lambda_1 = \sqrt{N - 1}$; $\lambda_1 = N - 1$

Figure 17.8: Changing connectivity and vulnerability of graphs with changing λ_1. A chain has the least connectivity, with $\lambda_1 \approx 2$, even for a chain of N nodes; an N-node star comes next, with $\lambda_1 = \sqrt{N - 1}$; and finally an N-node clique, with $\lambda_1 = N - 1$. Notice that the average degree labels the N-node star as equivalent to the N-node chain, focusing only on the one-step-away paths, while the eigenvalue takes into account paths of all lengths.

from the Oregon router views. It has 15,420 links and 3,995 nodes (AS peers)[1]. We use this dataset to illustrate the super-model theorem.

Figure 17.7 gives an overview of the simulations for the SIRS (whooping cough, limited-duration immunity) model. All values are average over several runs of the simulations. In short, as expected from the theorem, there is a qualitative difference of behavior when we are above (red), below (green), and at threshold.

Figure 17.7(A) shows a time-evolution plot of the fraction of infected nodes (*footprint*) in the graph, for different values of the effective strength of the virus. Specifically it gives results for *above threshold* (in red) and *under threshold* (in green).

We also give a "take-off" plot (Figure 17.7(B)). It shows the number of infections at steady state ('footprint'), versus the different strengths of the virus. If the infection resulted in an epidemic then the footprint will be large. As the theorem predicted, the plot shows that the tipping point is at the point when the effective strength $s = 1$.

17.4.2 λ_1: MEASURE OF CONNECTIVITY

What does the result mean exactly? Clearly, a graph that is better connected should be better for the virus. The most natural measure of connectivity is the average degree – why is it not enough? Why is it that the first eigenvalue is the determining factor?

Intuitively, λ_1 (also known as the spectral radius) of a graph captures the connectivity of the graph. It is superior to the average degree, for the following intuitive reasons:

- for "homogeneous"/"regular" graphs (i.e., where every node has the same degree d, reaching d random nodes), the eigenvalue is $\lambda_1 \approx d$, completely agreeing with intuition.
- for graphs with skewed degree distributions, like, say a "star," the average degree tells us how many nodes can reach other nodes, within just one step; the λ_1 takes more steps into account.

---

[1]See http://topology.eecs.umich.edu/data.html.

The intuition is best illustrated through the toy graphs of Figure 17.8. Although an N-node chain (A) and an N-node star (B) have the same count of edges and thus the same average degree, intuitively, the star has a better connectivity. The eigenvalue captures exactly that. In a chain, we have $\approx 2N$ pairs within one-step, and $\approx 4N$ pairs within two-steps; for a star, we still have $\approx 2N$ pairs within one-step, but $\approx N^2$ pairs within two steps. This is why the first eigenvalue is larger for the star. Very informally, here λ_1 can be seen as the square root of two step-away paths. Not surprisingly, for a clique of N nodes, the eigenvalue is higher than the chain and the star ($\lambda_1 = N - 1$), correctly reflecting the excellent connectivity of the clique.

In short, the summary of our intuitive discussion is:

Observation 17.2 As a measure of connectivity, the eigenvalue λ_1 is like the average degree, but averaged properly, to take into account all path-lengths.

And, of course, the better connected a graph is (high λ_1), the better it is for the virus.

17.5 CONCLUSION

The work in [232] provided two orthogonal generalizations of earlier epidemic threshold results.

- In the first direction, the result gives the threshold of the generalized $S^*I^2V^*$ model, which encompasses *any* epidemic model in published literature ([140], [105], etc.).
- In the second direction, topology, it showed that for any, arbitrary, undirected contact-network, the effect of the topology can be captured solely by λ_1, the first eigenvalue of the adjacency matrix.

Moreover, we discussed some important applications and implications of the result for policy makers, scientists, marketers, and specifically:

- Fast answers to "what-if" questions.
- Guiding immunization policies: immunize those nodes that drop the eigenvalue as much as possible.
- Simplifying epidemiological simulations: if we are below threshold, we don't need to do a simulation at all, since the virus will die out.

CHAPTER 18

Case Studies

18.1 PROXIMITY AND RANDOM WALKS

Motivation: Given a social network, as in Figure 18.1(a), and two persons (nodes) A and E, how close are they to each other, so that we can recommend that they become potential friends? How much influence does node i have on node j? How does this change if we take the strength of the ties into account (indicated as numerical values above the edges)?

Similarly, in a setting where user rate products (like, e.g., Amazon and Netflix), which product is best to recommend to user B of, say, Figure 18.1(b)?

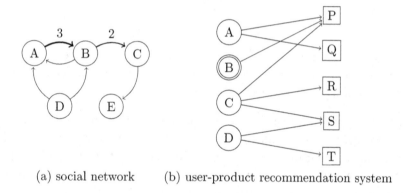

(a) social network   (b) user-product recommendation system

Figure 18.1: Examples where we need node-proximity: (a) a social network, where we want to judge how close is, say, "A" to "E." (b) a user-product recommendation system, where we want to recommend a new product to, say, user "B."

Intuition: Clearly, two nodes are close to each other if there are *several, short, heavy-weighted* paths between them. In Figure 18.1(a), suppose that the numbers indicate the strength of the tie (with the default weight, 1, omitted for clarity).

There are numerous measures, like shortest path and maximum flow, but they have shortcomings (see [113]). One of the successfully used measures is the metric based on the so-called RWR (for *random walk with restarts*), also known as *personalized random walk* [137]. The idea is to modify the PageRank algorithm we saw before in Section 14.4, by forcing the random surfer to teleport always to the same node.

In more detail, suppose we want to find the influence/proximity $s_A(E)$, that is, how close is node "E" to node "A." Notice that there is no symmetry, even on an undirected graph: $s_A(E)$ usually differs from $s_E(A)$). Intuitively, high $s_A(E)$ score means that, if node A knows a rumor (or is infected with some virus), then node E has high chances of learning about the rumor (or getting the virus), and higher than node, say, F, which has a lower $s_A(F)$ score.

Then we do the following thought experiment: Let a random surfer start from "A," and, with probability c, follow the edges (with preference proportional to their weight); with probability $1 - c$, the random surfer teleports back to node "A." After infinite such trials, we can estimate the steady-state probability that the surfer will find herself on node "E" – that is, exactly defined as the influence / proximity $s_A(E)$ of node "E" to node "A." Clearly, if there are short, numerous, heavy paths from "A" to "E," this probability will be high, agreeing with intuition.

Formulas: The equations are almost identical to the PageRank equations 14.8 with the only difference on the re-start vector. Using the same notation as in Section 14.4, the vector of proximities $\vec{s}_i = [s_{i,j}]^t$, (where $s_{i,j} = s_i(j)$) for destination nodes j (=1, ..., N), with respect to starting node i is given by

$$\vec{s}_i = 1/(1 - c)\,[\mathbf{I} - c\mathbf{B}]^{-1}\vec{e}_i \qquad (18.1)$$

where $\vec{e}_i = [0, 0, \ldots, 0, 1, 0, \ldots, 0]^T$ is a column vector with zeros everywhere, except at position i. As in section 14.4, $1 - c$ and $\mathbf{B}$ are defined as the fly-out probability, and as the to-from transition matrix (column-normalized), respectively.

Discussion: Node proximity has attracted a lot of interest, with several variations to the above basic theme: The standard measures, like the shortest path length and the maximum flow have subtle issues [113], like ignoring the count of alternative paths, or the length of the path, respectively. Some of the best alternatives include the sink-augmented delivered current [113, 222] cycle free effective conductance [175], personalized random walks with restart [139, 224] and variations [209], and many more [264].

Although not identical, these alternatives all behave according to intuition: they give higher scores when there are many, short, heavy-weighted paths, from source to destination.

With respect to applications, graph-node proximity is an important building block in many settings, like the so-called "connection subgraphs" [112]; "center-piece subgraphs" [265], [175]; personalized PageRank [137]; neighborhood and anomaly detection in bipartite graphs [255]; content-based image retrieval [139]; cross modal correlation discovery [224]; link prediction [192]; and "most related records" retrieval in relational databases, like in the BANKS system [6], the ObjectRank method [33], and the RelationalRank method [122].

Next we shall describe two of the many applications where the proximity/influence score is useful. The first is on automatic image captioning, and the second is on the so-called *center piece sub-graphs* (nodes that are central with respect to a set of seed nodes).

18.2 AUTOMATIC CAPTIONING—MULTI-MODAL QUERYING

Given a collection of multimedia objects, we want to find correlations across media. The driving application is auto-captioning, where the problem is defined as follows:

Problem 18.1 auto-captioning   Given a set S of color images, each with caption words; and given one more, uncaptioned image I, find the best t (say, $t=5$) caption words to assign to it.

See Figure 18.2(a-c) for an example. However, the upcoming *GCap* method is general, and can be applied to video clips (with text scripts, audio, motion); on audio songs, with text lyrics, and so on. We will refer to the following additional scenarios:

Scenario 1: Video auto-captioning For example, given a training set of captioned video clips, we want to explore the correlation between features of video clips and captions so that new unseen video clips can be captioned. The goal of this section is to seek associations between features and terms in captions.

Scenario 2: Multi-lingual text Given a collection of documents, in two languages, say, English and Spanish, find correlations between English and Spanish terms.

Scenario 3: Similarity search Given a collection of songs, with audio, lyrics, and singer names, ($m=3$ multimedia attributes), find songs similar to "Yesterday" Beatles – similar either in sound, or in terms of lyrics, or in terms of artists.

18.2.1 THE *GCAP* METHOD

Figure 18.2 gives a toy examples for the 3 sample images, and their captions. The last figure has no caption words, and our goal is to find the most suitable such words, from the existing ones. How can we turn this problem into a graph problem? We can easily map images into one type of nodes into another type of nodes with a link whenever an image has been assigned a term as a caption word. But we also need to capture somehow the image-to-image similarity.

The idea proposed in [225] is to introduce a third type of nodes "regions," and link similar regions. Specifically, we can divide all images into regions, extract features from these regions, and then link each region to similar regions. We will elaborate on the details and the design choices, but the summary is that eventually we have a graph like the one in Figure 18.2(g), and we want to find which of the "term" nodes ($t_1, \ldots t_8$) is closest to node I_3 – those terms are probably the best for captioning that image.

The original *GCap* paper [225] proposed to use any standard segmentation algorithms [246] for each image, break it into regions (see Figure 18.2(d,e,f)), and then map each region into a (say, 30-d) feature vector. The mapping can be done with any of the published feature extraction functions (color-histograms, texture histograms, etc. [111]). In that specific setting, we can extract $p=30$ features from each region ("blob"), like the mean and standard deviation of RGB values, average responses to various texture filters, its position in the entire image layout, and some shape

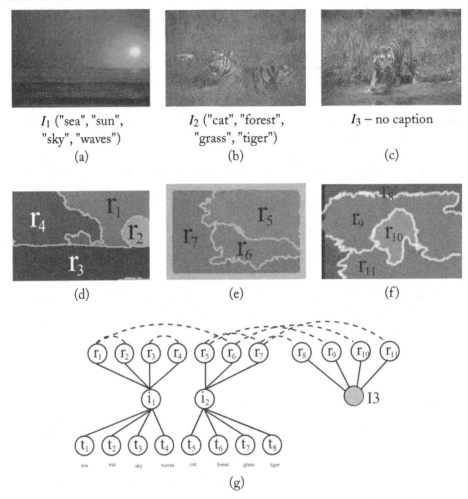

Figure 18.2: Three sample images, two of them annotated; their regions (d,e,f); and their *GCap* graph (g). (Figures look best in color.)

descriptors (e.g., major orientation and the area ratio of bounding region to the real region). Note that the exact feature extraction details are *orthogonal* to the approach – all that the *GCap* method needs is a black box that will map each color image into a set of zero or more feature vectors.

There are several subtle details that are discussed in [225]. For example, should we link all pairs of regions that have similarity above a threshold? How would one choose such a threshold? How can we quickly find all such pairs? The answer is to link each region to its *k* most similar regions, (with *k=3* being the recommended value, and the approximate nearest neighbor algorithm of Arya et al. [26] being the recommended algorithm).

Figure 18.2 illustrates the *GCap* approach with an example:

Example 1 *Consider the image set $\mathcal{S}=\mathcal{I}=\{I_1, I_2, I_3\}$ (Figure 18.2). The graph corresponds to this data set has three types of nodes: one for the image objects i_j's ($j = 1, 2, 3$); one for the regions r_j's ($j = 1, \ldots, 11$), and one for the terms $\{t_1, \ldots, t_8\}$={sea, sun, sky, waves, cat, forest, grass, tiger}. Figure 18.2(g) shows the resulting GCap graph. Solid arcs indicate Object–Attribute–Value relationships; dashed arcs indicate nearest–neighbor (NN) relationships.*

In this example, we consider only k=1 nearest neighbor, to avoid cluttering the diagram.

For example, to solve the auto-captioning problem for image I_3 of Figure 18.2, we can estimate the steady-state probabilities $u_{i3}(*)$ for all nodes of the *GCap*, we can keep only the nodes that correspond to terms, and we can report the top few (say, five), as caption words.

18.2.2 PERFORMANCE AND VARIATIONS

The original *GCap* method as reported in [225] shows good classification accuracy on standard image datasets, and is linear on the number of edges on query time. The graph construction time depends on how quickly we can find nearest neighbors. Approximations to this issue, and several variations, were studied in the so-called "QMAS" method [86], and we present results on satellite image labeling from there.

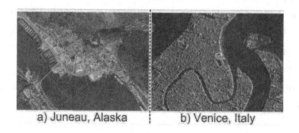

a) Juneau, Alaska   b) Venice, Italy

Figure 18.3: Examples of satellite images (Juneau, Alaska, USA; and Venice, Italy). Images are split in tiles, say, 16x16 pixels, and *some, few* of them are labeled ("water,' "concrete"). We want to label all the remaining ones.

Figure 18.3 gives some sample satellite images that were used in the "QMAS" method [86, 135]. In this setting, the images are divided into tiles (say, square, 16x16 pixel tiles), and human experts label some of them by hand (e.g., "water," "concrete," "grass," "boat"). The goal is to automatically label the remaining (vast majority) of tiles. Of course, as with *GCap*, we need a domain expert to design a similarity function between two tiles.

Figure 18.4 shows the results of the "QMAS" method where it automatically labeled several tiles as "boats" (bottom row), after it was given only a handful of such labeled tiles (top row).

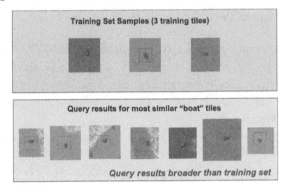

Figure 18.4: Result of labeling: top tiles are manually labeled as "boats;" "QMAS" returns several similar ones.

18.2.3 DISCUSSION

We presented *GCap* for image captioning, but the graph-based viewpoint can be applied to all the scenarios in the beginning of the section, with any combination of modalities we may have (music, voice, images, video, DNA-strings, etc.). The only requirement is that a domain expert would need to provide a similarity function for each modality.

Once we have turned the problem into a graph, as we did in Figure 18.2, then we can answer not only captioning questions (linking images to keywords), but questions from any modality to any other (images to images, keywords to regions, etc.). For example, in the above figure, we could ask "for region $r6$, what is the keyword that appears most often?" But we could ask even more complicated, mining, questions like clusters and anomalies. For example, "are there any major clusters?" (= graph partitions, in Figure 18.2(g)); "are there strange/rare regions, that look like nothing else?"

18.2.4 CONCLUSIONS

The graph-based view point of *GCap* and "QMAS" illustrates the power of graphs and node-proximity, in the unexpected setting of multimedia mining and querying. As long as we are given similarity functions between the involved modalities (images, video-clips, etc.), *GCap*, and random walks can handle numerous mining and querying tasks.

18.3 CENTER-PIECE SUBGRAPHS—WHO IS THE MASTERMIND?

Consider the social network of Figure 18.5, and suppose that the three gray nodes are criminals, while edges indicate friendship. Who is the central person (if any), of this criminal ring?

There are several more applications, in addition to law enforcement. For example, meme tracking – if nodes A, B, and C are researchers, and they are using a specific term or methodology, who could possibly have originated that, and how did the propagation take place?

In all cases, the goal is to find a few central nodes, and the connections to the given "query" nodes (A,B,C in our example). Using human intuition, in our toy example the best node is the red node in the center, with the pink nodes tying in second place.

The Connection Subgraph algorithm [113] addressed the case for $Q=2$ query nodes, and the Center-Piece Subgraph (CePS) algorithm [266] generalized to arbitrary number of Q nodes. Next we describe the latter.

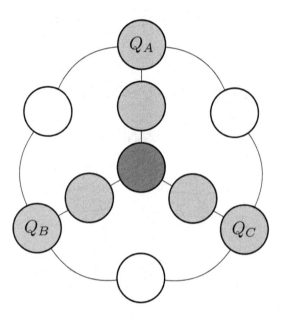

Figure 18.5: Example of *CEPS*: The gray nodes Q_A, Q_B, Q_C are the query nodes; the center node, in bright red, is the one best connected to all three.

Figure 18.6(a) gives screenshots of the *CEPS* system, showing the solution on a co-authorship graph from DBLP, with $Q=4$ query nodes. All four researchers are in data mining, but the first two (Rakesh Agrawal and Jiawei Han) are more on the database side, while Michael Jordan and Vladimir Vapnik are more on the machine learning and statistical side.

The results make sense: researchers like Daryl Pregibon, Padhraic Smythe, and Heikki Mannila are vital links, because of their strong cross-disciplinary connections with both of the above sub-areas.

The *CEPS* algorithm goes beyond plain conjunction ("AND"), allowing for the so-called "soft-ANDs." In a nutshell, in a $K\_softAND$ query, *CEPS* finds nodes with connections to at least

k of the query nodes, but not necessarily all of them (see Figure 18.6(b), which gives the results for $k = 2$ in the same DBLP co-authorship graph). Notice now that there are two disconnected groups, exactly reflecting the fact that our four query nodes come from two research areas, databases and machine-learning. Notice that the $K\_softAND$ query is very general: when $k=1$, it is identical to an "OR" query, while in the default case $(k = Q)$, it becomes an "AND" query.

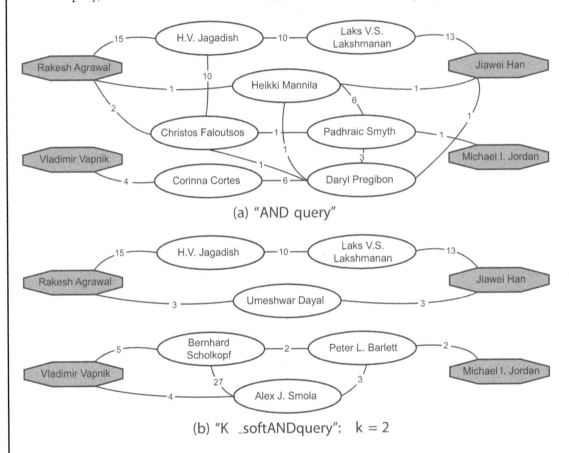

Figure 18.6: Center-piece subgraph among Rakesh Agrawal, Jiawei Han, Michael I. Jordan, and Vladimir Vapnik. Top: standard ("AND") query. Bottom: soft-AND query, requiring connections to at least $k=2$ of the query nodes. Notice that the top has weak connections, while the bottom is completely disconnected (data base/data mining community at top, and machine learning theory at the bottom).

Formally, the center-piece subgraph problem is defined as follows:

Problem 18.2   Center-Piece Subgraph Discovery(*CEPS*)

Given: an edge-weighted undirected graph $\mathbf{W}$, Q nodes as source queries $\mathcal{Q} = \{q_i\}$ $(i = 1, ..., Q)$, and an integer budget b.

Find: a suitably connected subgraph $\mathcal{H}$ that (a) contains all query nodes q_i, (b) at most b other vertices, and (c) it maximizes a "goodness" function $g(\mathcal{H})$.

Main idea As the reader may expect, the heart of the solution is in the node-proximity measure, where *CEPS* proposes to use personalized random walks with restarts. Node j obtains a high score with respect to query nodes $\mathcal{Q} = \{q_1, ..., q_Q\}$, if it is close to all of them, that is, if the *product* of its individual scores is high:

$$score(j, \mathcal{Q}) = \prod_{i=1}^{Q} s_{q_i}(j)$$

The product corresponds exactly to conjunction ("AND"). Intuitively, the score of node j is the probability that *all* Q surfers, each restarting from nodes $q_1, ..., q_Q$ respectively, will all find themselves on node j at steady state.

A $K\_softAND$ query is handled similarly – see the original paper for the details [266].

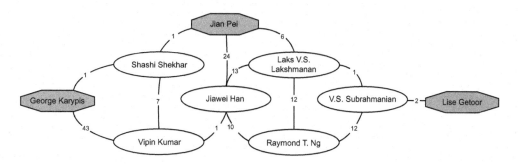

Figure 18.7: Example from DBLP

Results Figure 18.7 shows the result of *CEPS* when the query nodes are Profs. Lisa Getoor, Jian Pei, and George Karypis. Notice that the most central person seems to be Prof. Jiawei Han: he was the advisor of Prof. Jian Pei, had a collaboration with the Prof. Vipin Kumar (the advisor of Prof. Karypis), and collaborations with Prof. Raymond Ng, who had been advised by Prof. Subrahmanian at UMD. The latter is a colleague and collaborator of Prof. Getoor (also at UMD).

Discussion – Conclusions There are several elaborate fine-tunings in the algorithm: how to normalize the weights of the graph, how to find the shortest paths between the query nodes and the discovered nodes, how to pre-process the graph to speed up the computation (see [266]).

In conclusion, *CEPS* solves the "mastermind" problem, spotting nodes that are well connected to the given query nodes. It can also handle "soft-AND" queries, which are more general and more user-friendly. It is also based on the (personalized) random walk with restarts, illustrating the power of PageRank and its variations.

PART IV

Outreach—Related Work

CHAPTER 19

Social Networks

While the field of Graph Mining has been a relatively recent development in the Data Mining community, it has been studied under different guises by other groups of researchers, most notably by sociologists. Their work has culminated in the acceptance and usage of Social Network Analysis (SNA) as a tool for investigating the structure of social groups and organizations. For example, Cross et al. [88] use it to analyze the "invisible" patterns of interaction in organizations and map the informal networks in use. Weeks et al. [281] map the social network of drug users who exchange needles (and hence may spread AIDS).

Below, we give a very brief introduction to some of the important concepts. Interested readers are referred to the excellent book by Wasserman and Faust [277]. A nice introduction can also be found in [136]. Many of the concepts discussed in the previous sections also show up in Social Network Analysis, but under different names; Table 19.1 gives the meanings of some of them.

Table 19.1: *Social networks terminology:* We provide a list of typical terms used in Social Network Analysis, and their how they correspond to terms we are familiar with.

Term from Social Network Analysis	Meaning
Social network, or Sociogram	Graph
Actor	Node
Link	Edge
Ego	Current node under discussion
Alters	Other nodes, as viewed from the Ego
Bonacich centrality of a node	Abs. value of the first eigenvector component corresponding to that node
Betweenness centrality	"Stress" (see Definition 16.2)

19.1 MAPPING SOCIAL NETWORKS

One important aspect of SNA is getting the data itself. Unlike networks like the Internet, WWW, or metabolic pathways, social network mapping is not easily amenable to automated techniques. The primary method of explication involves interviewing the subjects. Formulation of interview questions is interesting in its own right, but not relevant to our work. Another question is: how do we choose the people to interview? There are two basic approaches:

- *Full network methods:* Here, we prespecify the set of actors, and then collect data on all the edges in the graph. This is similar to a census, and the final result is information about all the existing ties among the actors. Thus, the mapped social network is complete, but collecting the data is very hard.

- *Snowball methods:* We start with a focal actor or set of actors. We ask them to name their neighbors in the graph. These neighbors are added to the set, and we continue the process till some stopping criterion is reached. While this method is cheaper, it does not locate "isolated" actors, nor does it guarantee finding all connected individuals in the population. This method also tends to overstate the "solidarity" of the population.

19.2   DATASET CHARACTERISTICS

In general, the patterns we look for in SNA are similar to the ones discussed previously in Section 6. For each actor, we check his/her in-degree (related to his *prestige*), out-degree (related to his *influence*), and *distance* to other actors. For the entire network, we can calculate *size* (number of nodes), *density*, *reciprocity* (if I know you, what are the chances that you know me, too?) and *transitivity* (equivalent to the clustering coefficient). Freeman [119] defines *betweenness*, which we have already seen in previous sections. UCINET [56] is one well-known software implementing these.

An important aspect of SNA is the determination of an actor's *centrality* in the network. Many measures of centrality are in use, each having some distinct characteristics:

- *Degree centrality:* Nodes with a high degree are considered to be more central. However, this weights a node only by its immediate neighbors and not by, say, its two-hop and three-hop neighbors.

- *Bonacich centrality:* This measure uses the degree of the indirect neighbors, too [52]. The Bonacich centralities of the nodes in a graph are precisely the components of the first eigenvector of the graph.

- *Closeness centrality:* This is the (normalized) inverse of the average distance metric. Nodes which have low distance to all other nodes in the graph have high closeness centrality.

- *Betweenness centrality:* Nodes with high betweenness values occur on more "shortest-paths," and are presumably more important than nodes with low betweenness.

- *Flow centrality:* This is similar to betweenness centrality, except that instead of considering only the shortest paths between pairs of nodes, we consider all paths.

19.3   STRUCTURE FROM DATA

The classic "community structure" is a clique. However, the strict clique definition may be too strong for various applications. Several relaxed definitions have been proposed, such as:

- *N-clique:* Each node in an *N*-clique must be able to reach every other node in it within *N* hops. However, these paths may pass through non-members of the *N*-clique.

- *N-clan:* This is an *N*-clique with the restriction that all pairs of nodes in it should have a path of at most *N* hops passing only through other nodes of the *N*-clan.

- *K-plex:* Each member must have edges to all but *K* other members.

- *K-core:* Each member must have edges to *at least K* other members.

Another common notion in SNA is that of a core/periphery structure. Intuitively, a (sub)graph consists of a cohesive core with some sparse peripheral nodes. Borgatti and Everett [55] model an "ideal" core/periphery structure as an adjacency matrix with a block of 1 values for the core-core edges and a block of 0 values for the periphery-periphery edges. To actually find the core and periphery nodes in a given network, they use a function optimization routine which tries to maximize the correlation between the given graph and such an "ideal" graph. This, however, might not be easy or efficient for huge graphs.

19.4   SOCIAL "ROLES"

This refers to an abstract concept regarding the "position" of a actor in society. This could be based, in part, on the relationships that the actor in question has with other actors. For example, the "husband" role is defined in part as being linked to another actor in a "wife" role. In other words, social roles could be thought of as representing regularities in the relationships between actors.

Actors playing a particular social role have to be equivalent/similar to each other by some metric. In general, the following three kinds of similarities are considered, in decreasing order of constraints [54, 136]:

- *Structural equivalence: Two actors u and v in a graph $G = (V, E)$ are structurally equivalent iff [54]*

$$\forall \ actors \ x \in V, \quad (u, x) \in E \Leftrightarrow (v, x) \in E$$
$$and, \quad (x, u) \in E \Leftrightarrow (x, v) \in E$$

  In other words, they are linked to exactly the same set of nodes, with (in the case of directed graphs) the arrows pointing in the same directions. Two structurally equivalent actors can exchange their positions without changing the network. Figure 19.1(a) shows an example of this.

- *Automorphic equivalence: Two actors u and v of a labeled graph G are automorphically equivalent iff all the actors of G can be re-labeled to form and isomorphic graph with the labels of u and v interchanged [136].* Two automorphically equivalent vertices share exactly the same label-independent properties. For example, in Figure 19.1(b1), nodes *u* and *v* are not structurally

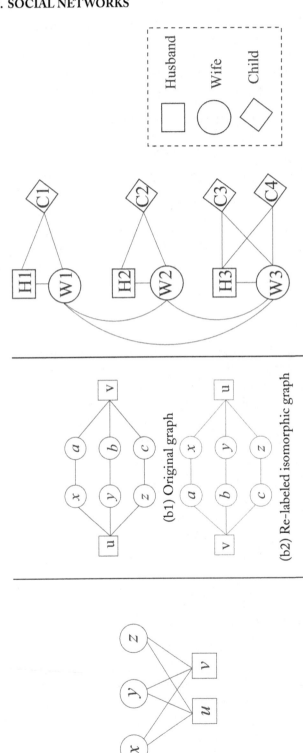

(a) Structural equivalence

(b) Automorphic equivalence

(c) Regular equivalence

(b1) Original graph

(b2) Re-labeled isomorphic graph

Husband

Wife

Child

Figure 19.1: *Social "roles" and equivalences:* (a) Nodes u and v are structurally equivalent because they are linked to exactly the same set of nodes: x, y, and z. (b) Nodes u and v of plot (b1) are automorphically equivalent because we can exchange their positions, and then re-label the rest of the nodes as in plot (b2), so that the new graph is isomorphic to the original. (c) The squares ("husbands"), circles ("wives") and rhombuses ("children") form three regularly equivalent classes. A "husband" connects to a "wife" and a "child;" a "wife" to a "husband," "child," and other "wives;" a "child" connects to a "husband" and a "wife." Thus, each class is defined by its set of relationships to the other classes. Note that child C1 and C4 are not structurally or automorphically equivalent, but they are regularly equivalent.

equivalent (u and v have neighbors with different labels). However, if we exchange their positions, we can re-label the rest of the nodes (by exchanging x and a, y and b, and z and c) such that the new labeled graph of Figure 19.1(b2) is isomorphic to the original. Thus, automorphic equivalence is a weaker condition than structural equivalence.

- *Regular equivalence: If $G = (V, E)$ is a connected graph and $\equiv$ is an equivalence relation on V, then $\equiv$ is a regular equivalence iff [54]*

$$\forall a, b, c \in V, a \equiv b \Leftrightarrow \begin{cases} \text{(i) } (a, c) \in E \Rightarrow \exists d \in V \text{ such that } (b, d) \in E \text{ and } d \equiv c \\ \text{(ii) } (c, a) \in E \Rightarrow \exists d \in V \text{ such that } (d, b) \in E \text{ and } d \equiv c \end{cases}$$

Two actors u and v are regularly equivalent if they are equally related to equivalent others; thus, the definition is recursive. Figure 19.1(c) shows an example with three regularly equivalent classes, each of which connects to a particular subset of classes (e.g., a "child" connects to a "husband" and a "wife.")

Structural equivalence has the strongest constraints, while regular equivalence has the weakest. However, regular equivalence is the hardest to compute, and is the equivalence of most interest to sociologists.

Practical computation of these equivalence classes can be computationally expensive [136], so the definitions are usually relaxed while analyzing real-world graphs. The following heuristics are often used:

- *Computing structural equivalence:* The correlation coefficient between actors is often used to measure the degree of structural equivalence between actors.

- *Computing automorphic equivalence:* Automorphic equivalence classes can be approximated using inter-actor distances: for each actor, we form a *sorted* vector of its distances to other actors. Two automorphic actors should have exactly the same distance profiles, and two "almost-automorphic" actors should have "similar" profiles. This idea is encapsulated in a heuristic which computes Euclidean distances between distance profiles, and clusters actors with low profile distances.

- *Computing regular equivalence:* Heuristics are used to compute some similarity measure between actors. However, irrespective of the similarity metric used, finding equivalent actors essentially reduces to a problem of clustering actors based on the (perhaps thresholded) similarity matrix. One such technique uses *Tabu search* [128] to group actors (forming blocks in the matrix) so that the variance within the blocks is minimized.

Finding such equivalence classes is a major goal of social network analysis, and any advances in algorithms or heuristics can have a major impact in this field.

19.5   SOCIAL CAPITAL

Social capital is essentially the idea that better connected people enjoy higher returns on their efforts. An individual occupying some special location in the social network might be in a position to broker information or facilitate the work of others or be important to others in some fashion. This importance could be leveraged to gain some profit. However, the problem is: what does "better connected" mean?

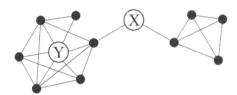

Figure 19.2: *Two concepts of social capital:* Node *X* has importance because it bridges the structural hole between the two clusters. Node *Y* is in the middle of a dense web, which provides easy access to reliable information; thus *Y* also has a good position in the network.

In general, there are two viewpoints on what generates social capital (Figure 19.2):

- *Structural holes:* Weak connections between groups are holes in the social structure, and create an advantage for individuals whose relationships span the holes [63]. Such individuals get lots of brokerage opportunities, and can control the flow of information between groups to their benefit.

- *Network closure:* This is the view that networks with lots of connections are the source of social capital [82]. When the social network around an actor *Y* is dense, it means that information flow to *Y* is quick and usually reliable. Also, the high density means that no one around *Y* can escape the notice of others; hence, everyone is forced to be trustworthy (or face losing reputation). Thus, it is less risky for *Y* to trust others, and this can be beneficial to him/her.

Burt [64] finds that these two viewpoints might not be totally at odds with each other. If a group has high closure but low contacts across holes, the group is cohesive but has only one perspective/skill. Low closure but high contacts across holes leads to a disintegrated group of diverse perspectives. The best performance is achieved when both are high. Thus, structural holes and network closure supplement each other.

19.6   RECENT RESEARCH DIRECTIONS

Social Network Analysis has been used to analyze many networks, from organizational to networks of drug use to networks of friendship in schools, and many others. Now, SNA is moving in new directions, some of which are discussed below.

Terrorist and covert networks   Recent events have focused attention on the mapping and analysis of terrorist networks. There are several problems unique to this setting [251], primarily due to lack of information:

- *Incompleteness* due to missing nodes and edges.

- *Fuzzy boundaries* due to not knowing whom to include or exclude form the mapped network.

- *Dynamism* of the network due to changing edges and the strengths of association of the nodes.

Baker and Faulkner [32] find that while the need for efficiency drives the structure of legal social networks, secrecy is the paramount concern in illegal networks. When the information processing needs are low, it leads to decentralized structures which protect the "ringleaders." However, for high information processing situations, the leaders must necessarily be at the core of the illegal network, increasing their vulnerability.

Krebs [177] tries to map the social network of the September-11 hijackers and some of their contacts, using only public data sources. He finds a very sparse network where many hijackers on the same team were far away from each other on the network. Coordination between far-off nodes is achieved via shortcuts in the network (as in the small-world model of Watts and Strogatz [279]). Trusted prior contacts kept the cells interconnected. Dombroski et al. [94] use the high-clustering coefficient and other properties of typical networks to estimate missing information in covert networks.

The Key Player problem   Who are the *most important* actors in a given social network? Borgatti [53] defines two "key-player" problems:

- (KPP-1) Find a set of k nodes whose removal maximally disrupts/disconnects the network. These individuals might be targeted for immunization to prevent an infection from becoming an epidemic.

- (KPP-2) Find a set of k nodes which are maximally connected to the rest of the network. These individuals could be targeted to diffuse information in a social network in the shortest possible time.

The "importances" of nodes are related, and choosing one node to be part of the top-k set changes the importances of others. Thus, finding the best set requires combinatorial optimization, which is very costly. Borgatti suggests using a greedy heuristic to solve this. However, formal error bounds on such heuristics are still not available.

19.7 DIFFERENCES FROM GRAPH MINING

We have seen in the previous paragraphs that Graph Mining and Social Network Analysis share many concepts and ideas. However, there are important differences too, the primary one being that of network size. Social networks are in general small, with the larger studies considering a few hundred nodes only. Graph Mining datasets, on the other hand, typically consist of hundreds of thousands of nodes and millions of edges. This difference in scale leads to many effects:

- *Power laws:* As we have seen in Section 6.1, power laws show up in a variety of situations in Graph Mining datasets, whereas they seem to be absent in almost all Social Network literature. This should be, in part, due to the fact that power-law patterns can be observed reliably only in large datasets.

- *Focus on computation costs:* For small networks, the existence of efficient algorithms is not an issue; even brute force algorithms can finish in a reasonably short time. However, some of the algorithms in use in SNA (such as combinatorial optimizations) become impractical for large datasets.

There has also been a difference in problems attracting research interest. The Graph Mining community does not seem to have worked on anything equivalent to social "roles" or the "power" of nodes. This might be due to the different semantics of the datasets in use. However, as newer and larger social network datasets (such as the "who-trusts-whom" dataset from `epinions.com`, the "who-reads-whose-weblog" dataset from blogspace, or the "who-knows-whom" dataset from `friendster.com`) become available, we might see a confluence of research in these two communities.

CHAPTER 20

Other Related Work

Several topics are closely related to graph mining, but have a different focus. Relational learning looks at the graph formed by interlinked relations in a database, and attempts to find patterns in it. Studies of rumor or viral propagation in a network look for key properties of the network which determine the susceptibility to epidemics. New graph navigation algorithms try to devise graphs so that local routing decisions can lead to nearly optimal global routes. These and other issues are discussed below.

20.1 RELATIONAL LEARNING

Relational data mining has attracted a lot of research interest recently [104]. While traditional data mining looks for patterns in a single database table, relational mining also uses the structure of linkage between multiple tables/relations. For example, given a database of movies, actors, awards, and the labeled links between them (i.e., a graph), McGovern and Jensen [204] find the patterns associated with movies being nominated for awards. The patterns themselves can be described as subgraphs with some relations as the nodes and some links between these relations as the edges.

Finding such patterns involves searching through a space of possible hypotheses, and we can do this search exhaustively or heuristically. One essential ingredient is the *pruning* of search paths which are not expected to lead to the solution. A widely used technique is Inductive Logic Programming (ILP), where patterns are expressed as logic programs. An advantage of this approach is the ability to easily incorporate background knowledge specific to the problem in the form of logic clauses. An alternative technique involves converting the relational data into a flat propositional format and then using well-known data mining tools on this flat relation; however, this conversion is non-trivial. In both cases, we run into efficiency concerns: the space of possible hypotheses is much larger than the case when we have a single relation, and searching in this space can be very costly. For example, even checking for the validity of a clause in ILP can be a costly affair, and solving a logic program typically involves checking the validity of many clauses.

Relational learning is a broad topic, and its details are beyond the scope of this work (see [123] for a survey, and [124] for an introductory book on the subject). The main difference from Graph Mining is that the latter focuses primarily on global topological structure and properties of a graph, (for example, the distribution of eigenvalues of the entire graph), while relational mining typically tries to identify local structures in labeled nodes and edges, where the semantics of these labels are particularly important. However, as both fields mature, we see a growing confluence in both the problems being solved and the techniques being used.

20.2   FINDING FREQUENT SUBGRAPHS

The mining of frequent patterns was first introduced in a databases context by Agrawal and Srikant [8], and is possibly one of the most popular data mining techniques. Recently, these ideas have been extended and applied to large graph datasets, to find the most common patterns or "motifs" hidden in the graph structure, to compress the graph dataset, and for many other problems. Below, we will discuss several of these methods.

APRIORI-like algorithms   Frequently occurring subgraphs in a large graph or a set of graphs could represent important motifs in the data. However, finding such motifs involves solving the graph and subgraph isomorphism problems, for which efficient solutions are not known (and subgraph isomorphism is known to be NP-complete). Most algorithms follow the general principle of the Apriori algorithm [8] for association rule mining. Inokuchi et al. [147] develop an Apriori-inspired algorithm called AGM, where they find a "canonical code" for any adjacency matrix and use these canonical codes for subgraph matching. However, this suffers from computational intractability when the graph size becomes too large. Kuramochi and Karypis [180] propose the FSG algorithm which also has the same flavor: starting with frequent subgraphs of one and two nodes, it successively generates larger subgraphs which still occur frequently in the graph. The algorithm expects a graph with colored edges and nodes; our graphs are a special case where all nodes and edges have only one color. However, it also needs to solve the graph and subgraph isomorphism problems repeatedly, and this is very slow and inefficient for graphs with only one color. Yan and Han [285] propose using a different canonical code based on depth-first search on subgraphs, and report faster results using this coding scheme.

On graphs where vertex coordinates are available, a more constrained version of this problem requires finding frequent *geometric* subgraphs. In this case, subgraph matching involves both topological and layout matching. Kuramochi and Karypis [181] find an algorithm that finds frequent geometric subgraphs that can be rotation, scaling, and translation invariant. This extra constraint allows the algorithm to finish in polynomial time.

"Difference from random" algorithm   Milo et al. [208] use another approach to find "interesting" motifs in a given graph. They define motifs to be "patterns that recur more frequently [*in a statistically significant sense*] in the real network than in an ensemble of randomized networks" (italics added). Each randomized network is generated so that each node in it has the same in-degree and out-degree as the corresponding node in the real network. However, this method assumes that matching the in- and out-degrees of a graph gives a good model of the graph; the motifs found under this assumption might not be statistically frequent if we used a better graph model.

Greedy algorithm   Holder et al. [144] try to solve a related but slightly different problem, that of compressing graphs using frequently occurring subgraphs. The subgraphs are chosen to minimize the minimum description length of the entire graph. They also allow *inexact matching* of subgraphs by assigning a cost to each "distortion" like deletion, insertion, or substitution of nodes and edges. However, to avoid the excessive computational overhead, they use a (suboptimal) greedy beam search.

Using Inductive Logic Programming (ILP) Instead of defining a subgraph as just a labeled graph topology, Dehaspe and Toivonen [92] use ILP to allow first order predicates in the description of frequent subgraphs. Their WARMR system lets many subgraphs have one succinct description, and can reveal a higher-order pattern than just the simple "propositional" subgraphs. However, this involves checking for equivalence of different first-order clauses, which is NP-complete. Nijssen and Kok [219] use a weaker equivalence condition in their FARMAR system to speed up the search. Still, finding first-order patterns is harder than finding propositional patterns, and it is unclear how fast such techniques will work on very large graph datasets.

20.3 NAVIGATION IN GRAPHS

The participants in Milgram's experiment [268] were able to build a chain to an unknown target individual, even though they knew only a few individuals in the full social network. We can navigate to websites containing the information we need, in spite of the immense size of the web. Such facts imply that large real-world graphs can be *navigated* with ease. In the following paragraphs, we will discuss methods of navigation that can be employed, and why real-world graphs seem to be so easy to navigate.

20.3.1 METHODS OF NAVIGATION

Some common methods for navigating large graphs include crawling, focused crawling, guided search based on power laws, and gossip protocols. We discuss each of these below.

Crawling: One question of interest in a graph is: given a starting node s in the graph, how can we reach some randomly assigned target node t? One technique involves having a search engine "crawl" the graph and store its data in a searchable form in some centralized system. Queries regarding the target node t can then be directed to this central server; this is the technique used by Web search engines like *Google* [59] or *CiteSeer* [126].

Focused crawling: How should we crawl pages while specifically looking for a particular topic? Chakrabarti et al. [74, 75, 76] propose a machine-learning approach to the problem using two "learners" working in tandem: (1) a "baseline learner" which can compute the degree of relevance of a given webpage to the required topic, and (2) an "apprentice learner" which computes the chances that a particular hyperlink points to a relevant webpage. Such a technique prevents the crawler from wasting effort on irrelevant portions of the Web; thus, the crawls can be conducted more frequently and the crawled data can be kept fresher.

Guided search using the power-law degree distribution: In case such a "directory" of nodes is not available, an alternative is to do a guided search. This involves making decisions based on local information only. Adamic et al. [5] use a message-passing system to search efficiently for some target data in the Gnutella network. The start node polls all its neighbors to see if any of them contains the required data, and if not, the search is forwarded to the neighbor with the highest degree. This neighbor now searches for the data among its neighbors, and so on. Full backtracking is implemented to prevent getting stuck in loops, and nodes do not get to see the same query twice. This technique is based on two ideas: (1) In scale-free networks, the path to a node of very high degree is usually short, and (2) the highest-connected nodes can (presumably) quickly spread the query to all other nodes in the network. However, this technique still requires $O(N)$ query messages to be sent between nodes, even though the path to the node containing the required data may be small.

Gossip protocols: Kempe et al. [159] take a different approach to search for "resources" in a network. They use a *gossip protocol* to spread information throughout a network about the availability of a "resource" at a particular node. Nodes share information with each other; the probability of communication between nodes u and v is a non-uniform inverse-polynomial function. However, this work assumes that any pair of nodes can communicate between themselves, and does not take the underlying graph structure into account.

20.3.2 RELATIONSHIP BETWEEN GRAPH TOPOLOGY AND EASE OF NAVIGATION

We will first discuss how a "good-for-navigation" graph can be designed, and then briefly touch upon some work investigating the reasons behind the ease of navigation on the WWW.

Designing "good-for-navigation" graphs Can we build a graph so that local routing decisions can be used to reach any target node via a short path? This is clearly what is happening in the social network of Milgram's experiment [207, 268]: not only does a short path exist between two randomly chosen people in the network, but such paths can also be *found* by the people, who forward the letters based only on the (local) information they have about the network structure. The problem has been studied in several forms, as described below.

2D Grid: Kleinberg [166] considers a model similar to that of Watts and Strogatz [279], but with a 2D lattice instead of a ring. Each node is connected to its neighbors in the lattice, but also has some *long-range* contacts; the probability of such a contact decreases exponentially with distance: $P(u, v) \propto d(u, v)^{-r}$. Based on the value of r, there are several cases:

- When $r = 2$, local routing decisions can lead to a path of expected length $O(\log N)$.

- When $0 \leq r < 2$, a local routing algorithm cannot use any "clues" provided by the geometry of the lattice, due to which path lengths are polynomial in N. This is in spite of the fact that (say for $r = 0$) there *exist* paths of length bounded by $\log N$ with high probability.

- When $r > 2$, long-range contacts are too few. Thus, the speed of moving toward the target node is too slow, leading to path lengths polynomial in N.

Thus, local routing decisions lead to a good path *only* when $r = 2$. The important points are twofold:

1. Minimizing the minimum expected number of steps from source to target is not necessarily the same as minimizing the diameter of the network.

2. In addition to having short paths, a network should also contain some latent structural clues to help make good routing decisions based on local information.

Hierarchies of attributes: Watts et al. [278] and (independently) Kleinberg [167] also proposed a different model to explain the goodness of local routing choices. The basic idea is that each person has a set of attributes, and is more likely to know people with similar attributes.

- Each person (node) has an individual *identity*, consisting of H attributes such as location, job, etc.

- Each attribute leads to a hierarchy/tree of nodes. For example, everyone in Boston is one cluster/leaf, everyone in NY is another leaf, and the Boston and NY leaves are closer by tree distance than, say, the Boston and LA leaves. Similar hierarchies exist for each attribute. Note that these hierarchies are *extra information*, and are unrelated to the social network itself.

- Two nodes u and v have a edge between them with a probability depending on how close they are in the attribute hierarchies. Specifically, let $d^a(u, v)$ be the height of the lowest common ancestor of nodes u and v in the hierarchy for attribute a. This measures the "distance" between the two nodes according to attribute a. We take the minimum distance over all hierarchies to be the "distance between u and v" $d(u, v)$. The probability of an edge (u, v) is exponential in this distance: $P(u, v) \propto e^{-\alpha d(u,v)}$.

The parameter α defines the structure of the social network; when it is large, we get isolated cliques, and the network looks more like a random graph as α increases. To pass on a message toward a prespecified target, everyone in the chain makes a local decision: he/she passes the letter to the node perceived to be closest to the target in terms of the minimum distance mentioned above.

Watts et al. observe that there is a wide range of α and H (the number of attributes) which lead to good routing decisions. In fact, they find that the best routing choices are made when $H = 2$ or 3, which agrees very well with certain sociological experiments [163]. Kleinberg [167] extends this to cases where we can form *groups* of individuals, which need not be hierarchies.

However, neither Watts et al. nor Kleinberg appear to consider the effect of power-law degree distributions in such graphs. Also, the probability of edge (u, v) is equal to that of the reverse edge (v, u); this might not hold always hold in practice.

Navigation in real-world graphs   In the context of social networks, Milgram's experiment [207, 268] shows the ability of people to choose "good" people to forward a letter to, so that the target can receive the letter in only a few steps. Dill et al. [93] find similar behavior in the WWW. In Section 5.2, we discussed the basic structure of the WWW, consisting of several subgraphs with one Strongly Connected Component (SCC) each. The authors find that the SCCs of the different subgraphs are actually very well connected between themselves, via the SCC of the entire WWW. This allows easy navigation of webpages: starting from a webpage, we progress to its SCC, travel via the SCC of the Web to the SCC of the target webpage, and from there onward to the target.

20.4   USING SOCIAL NETWORKS IN OTHER FIELDS

Viral Marketing   Traditional mass marketing techniques promote a product indiscriminately to all potential customers. Better than that is direct marketing, where we first attempt to select the most profitable customers to market to, and then market only to them. However, this only considers each person in isolation; the effects of one person's buying decisions on his/her neighbors in the social network are not considered. *Viral marketing* is based on the idea that considering social "word-of-mouth" effects might lower marketing costs.

Domingos and Richardson [95] model the problem as finding a boolean vector of whether to market to a person or not, such that the expected profit over the entire network is maximized. The key assumption is that each person's decisions on whether to buy a product or not are independent of the entire network, *given* the decisions of his/her neighbors. However, this formalism has poor scalability. The same authors make linearity assumptions to make the problem tractable for large graphs, and also allow the exact amount of discount offered to each person to be optimized [239]. The effect of a person on his neighbors might be mined from collaborative filtering systems or knowledge-sharing sites such as `www.epinions.com`. This is a very promising area of research, and is of prime interest to businesses.

Recommendation systems   Collaborative filtering systems have been widely used to provide recommendations to users based on previous information about their personal tastes. Examples include the *EachMovie* system to recommend movies (`www.research.compaq.com/src/eachmovie/`) and the product recommendation system of `www.amazon.com`. Such systems have been combined with social networks in the *ReferralWeb* system [158]. Given a particular topic, it tries to find an expert on that topic *who is related to the user by a short path in the social network*. This *path of referrals* might give the user a measure of trust in the recommended expert; it might also encourage the expert to answer the user's query.

CHAPTER 21

Conclusions

The focus of this book was real graphs, what they look like (Part I), how to mimic them with simple graph-generators (Part II), how to analyze them with some powerful tools from linear algebra (Part III), and a quick survey of concepts from related fields (Part IV). Here we list the major concepts from each part, and conclude with research directions.

Part I – Patterns   With respect to patterns, the most striking, recurring patterns in real graphs are the following:

- Power laws, in degree distributions, in PageRank distributions, in eigenvalue-versus-rank plots and many others.
- Small, and *shrinking* diameters, such as the "six degrees of separation" for the US social network, seven for the web [153], and four for Facebook [27]; even more, diameters seem to shrink as the graph grows [186].
- Triangle laws: there are many more triangles than random, and they even obey power laws: the more contacts a node has, the superlinearly more triangles it will participate in.
- Fortification laws, for weighted graphs: the more contacts one has, the super-linearly more phonecalls she will make.

Part II – Generators   The above patterns help us weed out generators. Of course, "all models are wrong," and all generators have weaknesses (and even future ones will always have weaknesses). However, we list the main ideas behind the most influential generators.

- *Preferential attachment:* Existing nodes with a high degree tend to attract more edges to themselves. This basic idea can lead to power-law degree distributions and small diameter.
- *"Copying" models:* Popular nodes get "copied" by new nodes, and this leads to power-law degree distributions as well as a community structure.
- *Constrained optimization:* Power laws can also result from optimizations of resource allocation under constraints.
- *Small-world models:* Each node connects to all of its "close" neighbors and a few "far-off" acquaintances. This can yield low diameters and high clustering coefficients.
- *Self-similarity:* Real graphs seem to have communities-within-communities, recursively. This is captured by the *RMat* model, and its derivatives ("Kronecker graphs"). Recursion is related to fractals, which naturally lead to "scale-free" properties and to power laws.

Part III – Tools   We discussed powerful tools for graph mining, and specifically SVD and related spectral concepts (eigenvalues, tensors). The main conclusions from there, are:

- *Ranking methods*: PageRank and HITS eventually reduce to eigenvectors or singular vectors.
- *Node proximity*: "Personalized Random Walks with Restarts" (RWR) lead to node-proximity functions that have numerous applications, like automatic image captioning.
- *Tensors*: an excellent tool for handling multi-aspect datasets, like graphs with edges of multiple types, and/or evolving over time.
- *Community detection*: we covered numerous graph partitioning methods, as well as the paradox of "no good cuts" that appears in numerous real networks.
- *Epidemic threshold*: We covered the G2 theorem, which shows that, for almost *any* virus type (rumor/product, etc.), the epidemic threshold (= tipping point between extinction and survival) depends on the first eigenvalue of the matrix.

Part IV – Outreach We covered concepts from sociology (social roles, social capital), from relational learning and from the literature of frequent subgraph discovery.

21.1 FUTURE RESEARCH DIRECTIONS

Although we covered a lot of material in this book, there are high-impact future directions, in three orthogonal groups:

Theory Static, unweighted graphs have attracted most of the research attention, because they are the simplest to find and study. However, there are fascinating, challenging extensions, with the cross-product of the following graph types:

- weighted graphs (when edges have weights, like count of phonecalls from "Alice" to "Bob");
- directed graphs (*if "Alice" calls "Bob" 20 times, how many times would "Bob" call "Alice?"*)
- dynamic, i.e., time-evolving graphs (*if "Alice" leaves the phone company, how many of her contacts will leave, too?*)
- nodes with attributes (*if we know that "Alice" is a 20-yr-old female, what can we guess about her contacts?*)
- edges with attributes, as in the tensor setting we saw in Section 15.1

Finally, a fascinating direction is how to *control* networks, like, e.g., protein-protein interaction networks, to disrupt cancer-like evolution in cells, or to block pathways that lead to Alzheimer's disease [194].

Applications Graph mining is a broad, multi-faceted challenge, with much room for innovation and discoveries. It benefits from, and also yields insights into, a breath taking spectrum of disciplines, including mathematics (matrix algebra, dynamical systems) [162], computer science (databases, machine learning, [7, 83, 288], parallel algorithms, and systems [155]), sociology [277], biology, medicine (brain wiring and the so-called "connectome" [134]), EEG studies and epilepsy detection and prediction [2, 77, 146]), economics (intelligent interacting agents [106]), marketing (product penetration with word-of-mouth reputation propagation [41, 95, 239]), ecology (with prey-predator relationships [197, 198]), epidemiology [24, 142], physics (cellular automata, "Ising spins," chaos theory [118, 243]), to name a few.

Engineering – Scalability   In the last few years, graph mining researchers have been extremely lucky: Large graph datasets are easier than ever to collect and work on (occasionally with a "non-disclosure agreement"), either by crawling the Web and social networks, or by collaborating with colleagues in large research centers (Google, Yahoo, MSR, Facebook). The volume of data is staggering, reaching billions of nodes in the published literature [183], [151], [153], and spanning multiple TeraBytes and possibly PetaBytes in large Internet/telecommunications companies and government.

The good news continues: not only data are available, but hardware keeps getting cheaper, bigger, and faster (about US$100 for a terabyte of disk, at the time of writing). On top of that, software for handling large distributed/parallel datasets is available, often in open source (e.g., *hadoop*[1], "pig,"[2] "pegasus"[3]).

The research challenges are how to re-think the typical graph algorithms, so that they scale up to the unprecedented scales we have available, with millions and billions of nodes.

21.2   PARTING THOUGHTS

The main high-level conclusion is that social networks, and graphs, in general, have attracted a lot of attention, thanks to their numerous, high impact *applications*, like recommendation systems, cyber-security, influence/product/virus propagation, to name a few.

The second high-level point is that graph mining is inherently cross-disciplinary, with fascinating opportunities for collaborations with all the disciplines we mentioned a few paragraphs above, and many more.

For these reasons, we believe that graph mining will be an active area of research for several years to come, and that it will produce fascinating discoveries, in numerous scientific fields – possibly, with some discoveries by you, dear reader!

---

[1]http://hadoop.apache.org/
[2]http://pig.apache.org/
[3]http://www.cs.cmu.edu/~pegasus/

Resources

There are numerous resources for graph mining, generators and related analyses. We list the ones we found most useful, sorted in our (subjective) order of usefulness. The provided URLs are active, at the moment of writing.

Datasets

1. `snap.stanford.edu/data`: The SNAP collection, by Prof. Jure Leskovec, with over 100 datasets and open source code.
2. `www-personal.umich.edu/~mejn/netdata`: About 15 datasets, heavily used (including the famous "Zachary karate" books) on U.S. politics, etc. It also has numerous pointers to other datasets. From U. Michigan, by Prof. Mark Newman.
3. `kdl.cs.umass.edu/data`: About four datasets, from UMass, by Prof. David Jensen.

Generators

1. `www.graph500.org`: The Graph500 generator, based on *RMat*, probably the best-supported generator.
2. `www.cs.cmu.edu/~lakoglu/#rtg`: RTG (Random Typing Generator) [14], with subtle improvements over *RMat*.
3. `www.cs.bu.edu/brite`: The "Brite" generator, which used to be a popular one for Internet-like topologies. No longer supported.

Graph mining software   Lists open-source software.

1. `www.cs.cmu.edu/~pegasus`: The CMU "Pegasus" system. In Java. It needs *hadoop*, and it scales to billion-node graphs. One of the few systems that does *not* expect the graph to fit in memory – probably the only one, in this section.
2. `networkx.lanl:gov`: If you like Python, you will love "networkx." It has ready-made functions for a huge variety of graph operations: degree distributions, clustering coefficient, random graph generation, etc.
3. `www.sandia.gov/~tgkolda/TensorToolbox`: Very popular tensor toolkit. In Matlab.
4. `jung.sourceforge.net`: JUNG (Java Universal Network/Graph Framework). We heard great comments about it from Java expert colleagues. Seems like the equivalent of "networkx," for Java.

General purpose packages

1. `igraph.sourceforge.net/doc/R/00Index.html`: The "igraph" package, works within the extremely popular and robust "R" statistical package (`www.r-project.org`), and can also

be called from python, and C. "igraph" provides a similar, broad, functionality like "networkx" and "JUNG" provide.

Graph Layout

1. `http://www.graphviz.org`: GraphViz—one of the oldest graph-layout packages, often used by other open-sources packages to do the graph layout (like "networkx" above, or "R"). It is *not* interactive.

Graph Visualization and Interaction

1. `http://www.cytoscape.org`: "Cytoscape" is well supported and carefully designed. The focus is on molecular interaction networks, but it works great for other, simpler networks as well.
2. `pajek.imfm.si/doku.php`: "Pajek" (=spider, in Slovenian). It is mainly for MS windows, but, with "wine," it has been reported to run on Linux/Mac-OSX. Well supported, very user friendly, with several articles and books written about it, and several network datasets on their wiki, in "pajek" format.
3. `graphexploration.cond.org`: "Guess" graph interaction system. We heard good comments about it, but it is unclear how well it is supported.

Virus propagation – Agents

1. `www.eclipse.org/stem`: SpatioTemporal Epidemiological Model (STEM): Simulates virus propagation on networks. It includes *real* connectivity data, between U.S. counties, as well as between countries of the world.
2. `ccl.northwestern.edu/netlogo`: "NetLogo" allows simulation of multiple agents, interacting with each other according to user-defined rules. It has numerous pre-fabricated scenarios (v. 5.0.1), among which the most related are the ones under `Models Library / Sample Models / Networks` and ... `/ Social Science`.

Bibliography

[1] E. Acar, C. Aykut-Bingol, H. Bingol, R. Bro, and B. Yener. Multiway analysis of epilepsy tensors. *Bioinformatics*, 23(13):i10–i18, 2007. Cited on page(s) 108

[2] Evrim Acar, Daniel M. Dunlavy, Tamara G. Kolda, and Morten Mørup. Scalable tensor factorizations with missing data. In *SDM*, pages 701–712. SIAM, 2010. Cited on page(s) 162

[3] Lada A. Adamic and Bernardo A. Huberman. Power-law distribution of the World Wide Web. *Science*, 287:2115, 2000. Cited on page(s) 55, 62

[4] Lada A. Adamic and Bernardo A. Huberman. The Web's hidden order. *Communications of the ACM*, 44(9):55–60, 2001. Cited on page(s) 37

[5] Lada A. Adamic, Rajan M. Lukose, A. R. Puniyani, and Bernardo A. Huberman. Search in power-law networks. *Physical Review E*, 64(4):046135 1–8, 2001. Cited on page(s) 158

[6] B. Aditya, Gaurav Bhalotia, Soumen Chakrabarti, Arvind Hulgeri, Charuta Nakhe, and S. Sudarshan Parag. Banks: Browsing and keyword searching in relational databases. In *VLDB*, pages 1083–1086, 2002. Cited on page(s) 136

[7] Charu Aggarwal. *Social Network Data Analytics*. Springer, 2011. Cited on page(s) 162

[8] Rakesh Agrawal and Ramakrishnan Srikant. Fast algorithms for mining association rules. In *Proc. of Intl. Conf. on Very Large Data Bases*, San Francisco, CA, 1994. Morgan Kaufmann. Cited on page(s) 98, 156

[9] William Aiello, Fan Chung, and Linyuan Lu. A random graph model for massive graphs. In *Proc. ACM SIGACT Symp. on the Theory of Computing*, pages 171–180, New York, NY, 2000. ACM Press. Cited on page(s) 47, 48, 51, 52, 61

[10] William Aiello, Fan Chung, and Linyuan Lu. Random evolution in massive graphs. In *IEEE Symposium on Foundations of Computer Science*, Los Alamitos, CA, 2001. IEEE Computer Society Press. Cited on page(s) 47, 48, 52, 60, 62

[11] L. Akoglu, M. McGlohon, and C. Faloutsos. RTM: Laws and a recursive generator for weighted time-evolving graphs. *Carnegie Mellon University Technical Report*, Oct, 2008. Cited on page(s) 24

[12] Leman Akoglu, Pedro O. S. Vaz de Melo, and Christos Faloutsos. Quantifying reciprocity in large weighted communication networks. In *PAKDD (2)*, pages 85–96, 2012. Cited on page(s) 42

[13] Leman Akoglu and Christos Faloutsos. RTG: A recursive realistic graph generator using random typing. In *Machine Learning and Knowledge Discovery in Databases, European Conference, ECML PKDD 2009, Bled, Slovenia, September 7-11, 2009, Proceedings, Part I*, pages 13–28, 2009. Cited on page(s) 89

[14] Leman Akoglu and Christos Faloutsos. RTG: a recursive realistic graph generator using random typing. *Data Min. Knowl. Discov.*, 19(2):194–209, 2009. Cited on page(s) 165

[15] Leman Akoglu, Mary McGlohon, and Christos Faloutsos. Rtm: Laws and a recursive generator for weighted time-evolving graphs. In *Proceedings of the 8th IEEE International Conference on Data Mining (ICDM 2008), December 15-19, 2008, Pisa, Italy*, pages 701–706, 2008. Cited on page(s) 89

[16] Réka Albert and Albert-László Barabási. Topology of evolving networks: local events and universality. *Physical Review Letters*, 85(24):5234–5237, 2000. Cited on page(s) 47, 48, 57, 60

[17] Réka Albert and Albert-László Barabási. Statistical mechanics of complex networks. *Reviews of Modern Physics*, 74(1):47–97, 2002. Cited on page(s) 15, 51, 52

[18] Reka Albert, Hawoong Jeong, and Albert-Laszlo Barabasi. Diameter of the world wide web. *Nature*, 401:130–131, 1999. Cited on page(s) 12

[19] Réka Albert, Hawoong Jeong, and Albert-László Barabási. Diameter of the World-Wide Web. *Nature*, 401:130–131, September 1999. Cited on page(s) 15

[20] Réka Albert, Hawoong Jeong, and Albert-László Barabási. Error and attack tolerance of complex networks. *Nature*, 406:378–381, 2000. Cited on page(s) 73, 77, 78, 130

[21] Noga Alon. Spectral techniques in graph algorithms. In C. L. Lucchesi and A. V. Moura, editors, *Lecture Notes in Computer Science 1380*, pages 206–215. Springer-Verlag, Berlin, 1998. Cited on page(s) 117

[22] Noga Alon, Raphy Yuster, and Uri Zwick. Finding and counting given length cycles. *Algorithmica*, 17(3):209–223, 1997. Cited on page(s) 114

[23] Luís A. Nunes Amaral, Antonio Scala, Marc Barthélémy, and H. Eugene Stanley. Classes of small-world networks. *Proceedings of the National Academy of Sciences*, 97(21):11149–11152, 2000. Cited on page(s) 37

[24] R. M. Anderson and R. M. May. Coevolution of hosts and parasites. *Parasitology*, 85, 1982. Cited on page(s) 163

[25] Roy M. Anderson and Robert M. May. *Infectious Diseases of Humans*. Oxford University Press, 1991. Cited on page(s) 128, 129

[26] S. Arya, D.M. Mount, N.S. Netanyahu, R. Silverman, and A.Y. Wu. An optimal algorithm for approximate nearest neighbor searching in fixed dimensions. *JACM*, 45(6):891–923, 1998. Cited on page(s) 138

[27] Lars Backstrom, Paolo Boldi, Marco Rosa, Johan Ugander, and Sebastiano Vigna. Four degrees of separation. In *WebSci*, 2012. Cited on page(s) 15, 161

[28] Brett W. Bader and Tamara G. Kolda. Efficient MATLAB computations with sparse and factored tensors. *SIAM Journal on Scientific Computing*, 30(1):205–231, December 2007. Cited on page(s) 110

[29] B.W. Bader, R.A. Harshman, and T.G. Kolda. Temporal analysis of social networks using three-way dedicom. *Sandia National Laboratories TR SAND2006-2161*, 2006. Cited on page(s) 108

[30] B.W. Bader and T.G. Kolda. Matlab tensor toolbox version 2.2. *Albuquerque, NM, USA: Sandia National Laboratories*, 2007. Cited on page(s) 110

[31] Ricardo Baeza-Yates and Barbara Poblete. Evolution of the Chilean Web structure composition. In *Latin American Web Congress*, Los Alamitos, CA, 2003. IEEE Computer Society Press. Cited on page(s) 32

[32] Wayne E. Baker and Robert R. Faulkner. The social organization of conspiracy: Illegal networks in the Heavy Electrical Equipment industry. *American Sociological Review*, 58(6):837–860, 1993. Cited on page(s) 153

[33] Andrey Balmin, Vagelis Hristidis, and Yannis Papakonstantinou. Objectrank: Authority-based keyword search in databases. In *VLDB*, pages 564–575, 2004. Cited on page(s) 136

[34] Ziv Bar-Yossef, Ravi Kumar, and D. Sivakumar. Reductions in streaming algorithms, with an application to counting triangles in graphs. In *ACM-SIAM Symposium on Discrete Algorithms*, Philadelphia, PA, 2002. SIAM. Cited on page(s) 114

[35] A. L. Barabasi and R. Albert. Emergence of scaling in random networks. *Science*, 286(5439):509–512, October 1999. Cited on page(s) 4

[36] Albert-Laszlo Barabasi. *Linked: How Everything Is Connected to Everything Else and What It Means for Business, Science, and Everyday Life*. Plume Books, April 2003. Cited on page(s) 12

[37] Albert-László Barabási and Réka Albert. Emergence of scaling in random networks. *Science*, 286:509–512, 1999. Cited on page(s) 37, 47, 48, 53

[38] Albert-László Barabási, Hawoong Jeong, Z. Néda, Erzsébet Ravasz, A. Schubert, and Tamás Vicsek. Evolution of the social network of scientific collaborations. *Physica A*, 311:590–614, 2002. Cited on page(s) 47, 48, 55, 61

[39] Michael Barnathan, Vasileios Megalooikonomou, Christos Faloutsos, Feroze B. Mohamed, and Scott Faro. High-order concept discovery in functional brain imaging. In *ISBI*, pages 664–667, 2010. Cited on page(s) 108

[40] A. Barrat, M. Barthélemy, and A. Vespignani. *Dynamical Processes on Complex Networks*. Cambridge University Press, 2010. Cited on page(s) 127

[41] Frank M. Bass. A new product growth for model consumer durables. *Management Science*, 15(5):215–227, 1969. Cited on page(s) 162

[42] Jan Beirlant, Tertius de Wet, and Yuri Goegebeur. A goodness-of-fit statistic for Pareto-type behaviour. *Journal of Computational and Applied Mathematics*, 186(1):99–116, 2005. Cited on page(s) 36

[43] Asa Ben-Hur and Isabelle Guyon. Detecting stable clusters using principal component analysis. In M. J. Brownstein and A. Khudorsky, editors, *Methods in Molecular Biology*, pages 159–182. Humana Press, Totowa, NJ, 2003. Cited on page(s) 117

[44] Noam Berger, Christian Borgs, Jennifer T. Chayes, Raissa M. D'Souza, and Bobby D. Kleinberg. Competition-induced preferential attachment. *Combinatorics, Probability and Computing*, 14:697–721, 2005. Cited on page(s) 74

[45] Michael W. Berry. Large scale singular value computations. *International Journal of Supercomputer Applications*, 6(1):13–49, 1992. Cited on page(s) 117

[46] Zhiqiang Bi, Christos Faloutsos, and Flip Korn. The DGX distribution for mining massive, skewed data. In *Conference of the ACM Special Interest Group on Knowledge Discovery and Data Mining*, pages 17–26, New York, NY, 2001. ACM Press. Cited on page(s) 37, 82, 84

[47] Ginestra Bianconi and Albert-László Barabási. Competition and multiscaling in evolving networks. *Europhysics Letters*, 54(4):436–442, 2001. Cited on page(s) 47, 48, 49, 56, 62

[48] Paolo Boldi, Bruno Codenotti, Massimo Santini, and Sebastiano Vigna. Structural properties of the African Web. In *International World Wide Web Conference*, New York, NY, 2002. ACM Press. Cited on page(s) 37

[49] Béla Bollobás. *Random Graphs*. Academic Press, London, 1985. Cited on page(s) 50

[50] Béla Bollobás, Christian Borgs, Jennifer T. Chayes, and Oliver Riordan. Directed scale-free graphs. In *ACM-SIAM Symposium on Discrete Algorithms*, Philadelphia, PA, 2003. SIAM. Cited on page(s) 62, 77, 78, 130

[51] Béla Bollobás and Oliver Riordan. The diameter of a scale-free random graph. Combinatorica, 2002. Cited on page(s) 55

[52] Phillip Bonacich. Power and centrality: a family of measures. *American Journal of Sociology*, 92(5):1170–1182, March 1987. Cited on page(s) 148

[53] Steve Borgatti. The key player problem. In *Proceedings of the National Academy of Sciences Workshop on Terrorism*, Washington DC, 2002. National Academy of Sciences. Cited on page(s) 153

[54] Steve Borgatti and Martin G. Everett. The class of all regular equivalences: algebraic structure and computation. *Social Networks*, 11:65–88, 1989. Cited on page(s) 149, 151

[55] Steve Borgatti and Martin G. Everett. Models of core/periphery structures. *Social Networks*, 21:275–295, 1999. Cited on page(s) 149

[56] Steve Borgatti, Martin G. Everett, and Linton C. Freeman. UCINET V User's Guide. Analytic Technologies, 1999. Cited on page(s) 148

[57] Christian Borgs, Jennifer T. Chayes, Mohammad Mahdian, and Amin Saberi. Exploring the community structure of newsgroups (Extended Abstract). In *Conference of the ACM Special Interest Group on Knowledge Discovery and Data Mining*, New York, NY, 2004. ACM Press. Cited on page(s) 113

[58] Ulrik Brandes, Marco Gaertler, and Dorothea Wagner. Experiments on graph clustering algorithms. In *European Symposium on Algorithms*, Berlin, Germany, 2003. Springer Verlag. Cited on page(s) 117

[59] Sergey Brin and Lawrence Page. The anatomy of a large-scale hypertextual Web search engine. *Computer Networks and ISDN Systems*, 30(1–7):107–117, 1998. Cited on page(s) 11, 63, 101, 103, 157

[60] R. Bro. Parafac. tutorial and applications. *Chemometrics and intelligent laboratory systems*, 38(2):149–171, 1997. Cited on page(s) 108

[61] Andrei Z. Broder, Ravi Kumar, Farzin Maghoul, Prabhakar Raghavan, Sridhar Rajagopalan, Raymie Stata, Andrew Tomkins, and Janet Wiener. Graph structure in the web: experiments and models. In *International World Wide Web Conference*, New York, NY, 2000. ACM Press. Cited on page(s) 15, 31, 33, 37

[62] Tian Bu and Don Towsley. On distinguishing between Internet power law topology generators. In *IEEE INFOCOM*, Los Alamitos, CA, 2002. IEEE Computer Society Press. Cited on page(s) 14, 47, 48, 60, 77, 78

[63] Ronald S. Burt. *Structural Holes*. Harvard University Press, Cambridge, MA, 1992. Cited on page(s) 152

[64] Ronald S. Burt. Structural holes versus network closure as social capital. In N. Lin, K. S. Cook, and Ronald S. Burt, editors, *Social Capital: Theory and Research*. Aldine de Gruyter, Hawthorne, NY, 2001. Cited on page(s) 152

[65] Kenneth L. Calvert, Matthew B. Doar, and Ellen W. Zegura. Modeling Internet topology. *IEEE Communications Magazine*, 35(6):160–163, 1997. Cited on page(s) 31, 75

[66] Jean M. Carlson and John Doyle. Highly optimized tolerance: A mechanism for power laws in designed systems. *Physical Review E*, 60(2):1412–1427, 1999. Cited on page(s) 72

[67] D. Chakrabarti, Y. Wang, C. Wang, J. Leskovec, and C. Faloutsos. Epidemic thresholds in real networks. *ACM TISSEC*, 10(4), 2008. Cited on page(s) 123, 127

[68] D. Chakrabarti, Y. Zhan, D. Blandford, C. Faloutsos, and G. Blelloch. Netmine: New mining tools for large graphs. In *SIAM-Data Mining Workshop on Link Analysis, Counter-terrorism and Privacy*, 2004. Cited on page(s) 120

[69] Deepayan Chakrabarti. AutoPart: Parameter-free graph partitioning and outlier detection. In *Conference on Principles and Practice of Knowledge Discovery in Databases*, Berlin, Germany, 2004. Springer. Cited on page(s) 118

[70] Deepayan Chakrabarti, Christos Faloutsos, and Yiping Zhan. Visualization of large networks with min-cut plots, a-plots and r-mat. *Int. J. Hum.-Comput. Stud.*, 65(5):434–445, 2007. Cited on page(s) 120

[71] Deepayan Chakrabarti, Spiros Papadimitriou, Dharmendra S. Modha, and Christos Faloutsos. Fully automatic Cross-associations. In *Conference of the ACM Special Interest Group on Knowledge Discovery and Data Mining*, New York, NY, 2004. ACM Press. Cited on page(s) 118

[72] Deepayan Chakrabarti, Yiping Zhan, and Christos Faloutsos. R-MAT: A recursive model for graph mining. *SIAM Int. Conf. on Data Mining*, April 2004. Cited on page(s) 10

[73] Deepayan Chakrabarti, Yiping Zhan, and Christos Faloutsos. R-MAT: A recursive model for graph mining. In *SIAM Data Mining Conference*, Philadelphia, PA, 2004. SIAM. Cited on page(s) 13, 47, 48, 81, 83

[74] Soumen Chakrabarti. Recent results in automatic Web resource discovery. *ACM Computing Surveys*, 31(4):17, 1999. Article Number 17. Cited on page(s) 157

[75] Soumen Chakrabarti, Kunal Punera, and Mallela Subramanyam. Accelerated focused crawling through online relevance feedback. In *International World Wide Web Conference*, pages 148–159, New York, NY, 2002. ACM Press. Cited on page(s) 157

[76] Soumen Chakrabarti, Martin van den Berg, and Byron E. Dom. Focused crawling: a new approach to topic-specific Web resource discovery. *Computer Networks*, 31(11–16):1623–1640, 1999. Cited on page(s) 157

[77] Wanpracha Art Chaovalitwongse, Rebecca S. Pottenger, Shouyi Wang, Ya-Ju Fan, and Leonidas D. Iasemidis. Pattern- and network-based classification techniques for multichannel medical data signals to improve brain diagnosis. *IEEE Transactions on Systems, Man, and Cybernetics, Part A*, 41(5):977–988, 2011. Cited on page(s) 162

[78] Q. Chen, H. Chang, Ramesh Govindan, Sugih Jamin, Scott Shenker, and Walter Willinger. The origin of power laws in Internet topologies revisited. In *IEEE INFOCOM*, Los Alamitos, CA, 2001. IEEE Computer Society Press. Cited on page(s) 57, 60, 61

[79] P.A. Chew, B.W. Bader, T.G. Kolda, and A. Abdelali. Cross-language information retrieval using parafac2. In *Proceedings of the 13th ACM SIGKDD international conference on Knowledge discovery and data mining*, pages 143–152. ACM, 2007. Cited on page(s) 108

[80] Fan Chung and Linyuan Lu. The average distances in random graphs with given expected degrees. *Proceedings of the National Academy of Sciences*, 99(25):15879–15882, 2002. Cited on page(s) 21

[81] Aaron Clauset, Mark E. J. Newman, and Christopher Moore. Finding community structure of very large networks. *Physical Review E*, 70:066111, 2004. Cited on page(s) 116

[82] James S. Coleman. Social capital in the creation of human capital. *American Journal of Sociology*, 94:S95–S121, 1988. Cited on page(s) 152

[83] Diane J. Cook and Lawrence B. Holder. *Mining Graph Data*. John Wiley and Sons, 2007. Cited on page(s) 162

[84] Colin Cooper and Alan Frieze. A general model of web graphs. *Random Structures and Algorithms*, 22(3):311–335, 2003. Cited on page(s) 63

[85] Colin Cooper and Alan Frieze. The size of the largest strongly connected component of a random digraph with a given degree sequence. *Combinatorics, Probability and Computing*, 13(3):319–337, 2004. Cited on page(s) 52

[86] Robson Leonardo Ferreira Cordeiro, Fan Guo, Donna S. Haverkamp, James H. Horne, Ellen K. Hughes, Gunhee Kim, Agma J. M. Traina, Caetano Traina Jr., and Christos Faloutsos. Qmas: Querying, mining and summarization of multi-modal databases. In *ICDM*, pages 785–790, 2010. Cited on page(s) 139

[87] Thomas H. Cormen, Charles E. Leiserson, and Ronald L. Rivest. *Introduction to algorithms*. MIT Press and McGraw-Hill Book Company, Cambridge, MA, 6th edition, 1992. Cited on page(s) 117

[88] Rob Cross, Steve Borgatti, and Andrew Parker. Making invisible work visible: Using social network analysis to support strategic collaboration. *California Management Review*, 44(2):25–46, 2002. Cited on page(s) 147

[89] Mark Crovella and Murad S. Taqqu. Estimating the heavy tail index from scaling properties. *Methodology and Computing in Applied Probability*, 1(1):55–79, 1999. Cited on page(s) 36

[90] Pedro O. S. Vaz de Melo, Leman Akoglu, Christos Faloutsos, and Antonio Alfredo Ferreira Loureiro. Surprising patterns for the call duration distribution of mobile phone users. In *ECML/PKDD (3)*, pages 354–369, 2010. Cited on page(s) 42

[91] Derek John de Solla Price. A general theory of bibliometric and other cumulative advantage processes. *Journal of the American Society for Information Science*, 27:292–306, 1976. Cited on page(s) 53

[92] Luc Dehaspe and Hannu Toivonen. Discovery of frequent datalog patterns. *Data Mining and Knowledge Discovery*, 3(1):7–36, 1999. Cited on page(s) 157

[93] Stephen Dill, Ravi Kumar, Kevin S. McCurley, Sridhar Rajagopalan, D. Sivakumar, and Andrew Tomkins. Self-similarity in the Web. In *Proc. of Intl. Conf. on Very Large Data Bases*, San Francisco, CA, 2001. Morgan Kaufmann. Cited on page(s) 32, 33, 160

[94] Matthew Dombroski, Paul Fischbeck, and Kathleen M. Carley. Estimating the shape of covert networks. In *Proceedings of the 8th International Command and Control Research and Technology Symposium*, 2003. Cited on page(s) 153

[95] Pedro Domingos and Matthew Richardson. Mining the network value of customers. In *Conference of the ACM Special Interest Group on Knowledge Discovery and Data Mining*, New York, NY, 2001. ACM Press. Cited on page(s) 11, 13, 160, 162

[96] Sergey N. Dorogovtsev, Alexander V. Goltsev, and José Fernando Mendes. Pseudofractal scale-free web. *Physical Review E*, 65(6):066122, 2002. Cited on page(s) 115

[97] Sergey N. Dorogovtsev and José Fernando Mendes. *Evolution of Networks: From Biological Nets to the Internet and WWW*. Oxford University Press, Oxford, UK, 2003. Cited on page(s) 47, 48, 56

[98] Sergey N. Dorogovtsev, José Fernando Mendes, and Alexander N. Samukhin. Structure of growing networks with preferential linking. *Physical Review Letters*, 85(21):4633–4636, 2000. Cited on page(s) 55, 56

[99] Sergey N. Dorogovtsev, José Fernando Mendes, and Alexander N. Samukhin. Giant strongly connected component of directed networks. *Physical Review E*, 64:025101 1–4, 2001. Cited on page(s) 48, 52, 55, 61

[100] John Doyle and Jean M. Carlson. Power laws, Highly Optimized Tolerance, and Generalized Source Coding. *Physical Review Letters*, 84(24):5656–5659, June 2000. Cited on page(s) 72

[101] Petros Drineas, Alan Frieze, Ravi Kannan, Santosh Vempala, and V. Vinay. Clustering in large graphs and matrices. In *ACM-SIAM Symposium on Discrete Algorithms*, Philadelphia, PA, 1999. SIAM. Cited on page(s) 117

[102] Nan Du, Christos Faloutsos, Bai Wang, and Leman Akoglu. Large human communication networks: patterns and a utility-driven generator. In *Proceedings of the 15th ACM SIGKDD International Conference on Knowledge Discovery and Data Mining, Paris, France, June 28 - July 1, 2009*, pages 269–278, 2009. Cited on page(s) 41

[103] Susan T. Dumais. Latent semantic indexing (LSI) and TREC-2. In D. K. Harman, editor, *The Second Text Retrieval Conference (TREC-2)*, pages 105–115, Gaithersburg, MD, Helen Martin 1994. NIST. Special publication 500-215. Cited on page(s) 98

[104] Sašo Džeroski and Nada Lavrač. *Relational Data Mining*. Springer Verlag, Berlin, Germany, 2001. Cited on page(s) 155

[105] D. Easley and J. Kleinberg. *Networks, Crowds, and Markets: Reasoning About a Highly Connected World*. Cambridge University Press, 2010. Cited on page(s) 123, 124, 128, 133

[106] David Easley and Jon Kleinberg. *Networks, Crowds, and Markets: Reasoning About a Highly Connected World*. Cambridge University Press, September 2010. Cited on page(s) 162

[107] Paul Erdős and Alfréd Rényi. On the evolution of random graphs. *Publication of the Mathematical Institute of the Hungarian Acadamy of Science*, 5:17–61, 1960. Cited on page(s) 46, 47, 48

[108] Paul Erdős and Alfréd Rényi. On the strength of connectedness of random graphs. *Acta Mathematica Scientia Hungary*, 12:261–267, 1961. Cited on page(s) 46

[109] Brian S. Everitt. *Cluster Analysis*. John Wiley, New York, NY, 1974. Cited on page(s) 115

[110] Alex Fabrikant, Elias Koutsoupias, and Christos H. Papadimitriou. Heuristically Optimized Trade-offs: A new paradigm for power laws in the Internet. In *International Colloquium on Automata, Languages and Programming*, pages 110–122, Berlin, Germany, 2002. Springer Verlag. Cited on page(s) 47, 48, 73

[111] Christos Faloutsos. *Searching Multimedia Databases by Content*. Kluwer, 1996. Cited on page(s) 137

[112] Christos Faloutsos, Kevin S. McCurley, and Andrew Tomkins. Fast discovery of connection subgraphs. In *KDD*, pages 118–127, 2004. Cited on page(s) 136

[113] Christos Faloutsos, Andrew Tomkins, and Kevin McCurley. Fast discovery of 'connection subgraphs'. *KDD*, August 2004. Cited on page(s) 135, 136, 141

[114] Michalis Faloutsos, Petros Faloutsos, and Christos Faloutsos. On power-law relationships of the internet topology. *SIGCOMM*, pages 251–262, Aug-Sept. 1999. Cited on page(s) 10, 41

[115] Michalis Faloutsos, Petros Faloutsos, and Christos Faloutsos. On power-law relationships of the Internet topology. In *Conference of the ACM Special Interest Group on Data Communications (SIGCOMM)*, pages 251–262, New York, NY, 1999. ACM Press. Cited on page(s) 11, 13, 15, 36

[116] Andrey Feuerverger and Peter Hall. Estimating a tail exponent by modelling departure from a Pareto distribution. *The Annals of Statistics*, 27(2):760–781, 1999. Cited on page(s) 36

[117] Gary W. Flake, Steve Lawrence, and C. Lee Giles. Efficient identification of Web communities. In *Conference of the ACM Special Interest Group on Knowledge Discovery and Data Mining*, New York, NY, 2000. ACM Press. Cited on page(s) 113, 116

[118] Gary Willian Flake. *The Computational Beauty of Nature - Computer Explorations of Fractals, Chaos, Complex Systems, and Adaptation*. The MIT Press, London, England, 1999. Cited on page(s) 163

[119] Linton C. Freeman. A set of measures of centrality based on betweenness. *Sociometry*, 40(1):35–41, 1977. Cited on page(s) 148

[120] A. Ganesh, L. Massoulie, and D. Towsley. The effect of network topology in spread of epidemics. *IEEE INFOCOM*, 2005. Cited on page(s) 123, 127

[121] D. Garlaschelli and M. I. Loffredo. Patterns of Link Reciprocity in Directed Networks. *Phys. Rev. Lett.*, 93:268701, 2004. Cited on page(s) 42

[122] Floris Geerts, Heikki Mannila, and Evimaria Terzi. Relational link-based ranking. In *VLDB*, pages 552–563, 2004. Cited on page(s) 136

[123] Lise Getoor and Christopher P. Diehl. Link mining: A survey. *SIGKDD Explorations*, 7(2):3–12, 2005. Cited on page(s) 155

[124] Lise Getoor and Ben Taskar, editors. *Introduction to Statistical Relational Learning*. MIT Press, 2007. Cited on page(s) 155

[125] David Gibson, Jon Kleinberg, and Prabhakar Raghavan. Inferring web communities from link topology. In *ACM Conference on Hypertext and Hypermedia*, pages 225–234, New York, NY, 1998. ACM Press. Cited on page(s) 117, 118

[126] C. Lee Giles, Kurt Bollacker, and Steve Lawrence. Citeseer: An automatic citation indexing system. In *ACM Conference on Digital Libraries*, New York, NY, 1998. ACM Press. Cited on page(s) 157

[127] Michelle Girvan and Mark E. J. Newman. Community structure in social and biological networks. In *Proceedings of the National Academy of Sciences*, volume 99, Washington DC, 2002. National Academy of Sciences. Cited on page(s) 116

[128] Fred Glover. Tabu search – part 1. *ORSA Journal on Computing*, 1(3):190–206, 1989. Cited on page(s) 151

[129] K.-I. Goh, E. Oh, Hawoong Jeong, B. Kahng, and D. Kim. Classificaton of scale-free networks. In *Proceedings of the National Academy of Sciences*, volume 99, pages 12583–12588, Washington DC, 2002. National Academy of Sciences. Cited on page(s) 116

[130] Michael L. Goldstein, Steven A. Morris, and Gary G. Yen. Problems with fitting to the power-law distribution. *The European Physics Journal B*, 41:255–258, 2004. Cited on page(s) 35, 36

[131] G. H. Golub and C. F. Van-Loan. *Matrix Computations*. The Johns Hopkins University Press, Baltimore, 2nd edition, 1989. Cited on page(s) 97

[132] Ramesh Govindan and Hongsuda Tangmunarunkit. Heuristics for Internet map discovery. In *IEEE INFOCOM*, pages 1371–1380, Los Alamitos, CA, March 2000. IEEE Computer Society Press. Cited on page(s) 15

[133] Mark S. Granovetter. The strength of weak ties. *The American Journal of Sociology*, 78(6):1360–1380, May 1973. Cited on page(s) 67

[134] William R. Gray, John A. Bogovic, Joshua T. Vogelstein, Bennett A. Landman, Jerry L. Prince, and R. Jacob Vogelstein. Magnetic resonance connectome automated pipeline. *IEEE EMBS*, 3(2):42–48, 2012. Also at arXiv:1111.2660v1. Cited on page(s) 162

[135] Fan Guo. *Mining and Querying Multimedia Data*. PhD thesis, CMU, 2011. CMU-CS-11-133. Cited on page(s) 139

[136] Robert A. Hanneman and Mark Riddle. Introduction to social network methods. http://faculty.ucr.edu/ hanneman/nettext/, 2005. Cited on page(s) 147, 149, 151

[137] Taher H. Haveliwala. Topic-sensitive pagerank. *WWW*, pages 517–526, 2002. Cited on page(s) 135, 136

[138] Yukio Hayashi, Masato Minoura, and Jun Matsukubo. Recoverable prevalence in growing scale-free networks and the effective immunization. *arXiv:cond-mat/0305549 v2*, Aug. 6 2003. Cited on page(s) 130

[139] Jingrui He, Mingjing Li, Hong-Jiang Zhang, Hanghang Tong, and Changshui Zhang. Manifold-ranking based image retrieval. In *ACM Multimedia*, pages 9–16, 2004. Cited on page(s) 136

[140] H. W. Hethcote. The mathematics of infectious diseases. *SIAM Review*, 42, 2000. Cited on page(s) 123, 124, 127, 128, 133

[141] H. W. Hethcote and J. A. Yorke. Gonorrhea transmission dynamics and control. *Springer Lecture Notes in Biomathematics*, 46, 1984. Cited on page(s) 123, 128

[142] Herbert W. Hethcote. The mathematics of infectious diseases. *SIAM Rev.*, 42(4):599–653, 2000. http://www.math.rutgers.edu/ leenheer/hethcote.pdf. Cited on page(s) 163

[143] Bruce M. Hill. A simple approach to inference about the tail of a distribution. *The Annals of Statistics*, 3(5):1163–1174, 1975. Cited on page(s) 36

[144] Lawrence Holder, Diane Cook, and Surnjani Djoko. Substructure discovery in the SUBDUE system. In *National Conference on Artificial Intelligence Workshop on Knowledge Discovery in Databases*, pages 169–180, Menlo Park, CA, 1994. AAAI Press. Cited on page(s) 156

[145] A. Hyvarinen, J. Karhunen, and E. Oja. *Independent Component Analysis*. John Wiley and Sons, 2001. Cited on page(s) 101

[146] L.D. Iasemidis, D.S. Shiau, J.C. Sackellares, P.M. Pardalos, and A. Prasad. A dynamical resetting of the human brain at epileptic seizures: Application of nonlinear dynamics and global optimization techniques. *IEEE T Bio-Med Eng*, 51(3):493–506, 2004. Cited on page(s) 162

[147] Akihiro Inokuchi, Takashi Washio, and Hiroshi Motoda. An apriori-based algorithm for mining frequent substructures from graph data. In *Conference on Principles and Practice of Knowledge Discovery in Databases*, Berlin, Germany, 2000. Springer. Cited on page(s) 156

[148] Hawoong Jeong, B. Tombor, Réka Albert, Zoltán N. Oltvai, and Albert-László Barabási. The large-scale organization of metabolic networks. *Nature*, 407(6804):651–654, 2000. Cited on page(s) 78

[149] U. Kang, Mary McGlohon, Leman Akoglu, and Christos Faloutsos. Patterns on the connected components of terabyte-scale graphs. In *ICDM*, pages 875–880, 2010. Cited on page(s) 41

[150] U. Kang, Brendan Meeder, and Christos Faloutsos. Spectral analysis for billion-scale graphs: Discoveries and implementation. In *PAKDD (2)*, pages 13–25, 2011. Cited on page(s) 15

[151] U. Kang, Brendan Meeder, and Christos Faloutsos. Spectral analysis for billion-scale graphs: Discoveries and implementation. In *PAKDD (2)*, pages 13–25, 2011. Cited on page(s) 163

[152] U. Kang, Evangelos Papalexakis, Abhay Harpale, and Christos Faloutsos. Gigatensor: Scaling tensor analysis up by 100 times - algorithms and discoveries. In *KDD*, 2012. Cited on page(s) 111

[153] U Kang, Charalampos (Babis) Tsourakakis, Ana Paula Appel, Christos Faloutsos, and Jure Leskovec. Radius plots for mining tera-byte scale graphs: Algorithms, patterns, and observations. *SDM'2010*, April-May 2010. Cited on page(s) 19, 161, 163

[154] U. Kang, Charalampos E. Tsourakakis, Ana Paula Appel, Christos Faloutsos, and Jure Leskovec. Hadi: Mining radii of large graphs. *TKDD*, 5(2):8, 2011. Cited on page(s) 15

[155] U. Kang, Charalampos E. Tsourakakis, and Christos Faloutsos. Pegasus: A peta-scale graph mining system. In *ICDM*, pages 229–238, 2009. Cited on page(s) 162

[156] Ravi Kannan, Santosh Vempala, and Adrian Vetta. On clusterings – good, bad and spectral. In *IEEE Symposium on Foundations of Computer Science*, Los Alamitos, CA, 2000. IEEE Computer Society Press. Cited on page(s) 117

[157] George Karypis and Vipin Kumar. Multilevel algorithms for multi-constraint graph partitioning. Technical Report 98-019, University of Minnesota, 1998. Cited on page(s) 117

[158] Henry Kautz, Bart Selman, and Mehul Shah. Referralweb: Combining social networks and collaborative filtering. *Communications of the ACM*, 40(3):63–65, 1997. Cited on page(s) 160

[159] David Kempe, Jon Kleinberg, and Alan J. Demers. Spatial gossip and resource location protocols. In *Proc. ACM SIGACT Symp. on the Theory of Computing*, New York, NY, 2001. ACM Press. Cited on page(s) 158

[160] J. O. Kephart and S. R. White. Directed-graph epidemiological models of computer viruses. *IEEE Computer Society Symposium on Research in Security and Privacy*, 1991. Cited on page(s) 123, 128

[161] J. O. Kephart and S. R. White. Measuring and modeling computer virus prevalence. *IEEE Computer Society Symposium on Research in Security and Privacy*, 1993. Cited on page(s) 123, 128

[162] Jeremy Kepner and John Gilbert. *Graph Algorithms in the Language of Linear Algebra*. SIAM, 2011. Cited on page(s) 89, 162

[163] Peter D. Killworth and H. Russell Bernard. Reverse small world experiment. *Social Networks*, 1(2):103–210, 1978. Cited on page(s) 159

[164] Jingu Kim and Haesun Park. Toward faster nonnegative matrix factorization: A new algorithm and comparisons. In *ICDM*, pages 353–362, 2008. Cited on page(s) 101

[165] Jon Kleinberg. Authoritative sources in a hyperlinked environment. *J. ACM*, 46(5):604–632, 1999. Cited on page(s) 101, 102, 117

[166] Jon Kleinberg. The small-world phenomenon: an algorithmic perspective. Technical Report 99-1776, Cornell Computer Science Department, 1999. Cited on page(s) 158

[167] Jon Kleinberg. Small world phenomena and the dynamics of information. In *Neural Information Processing Systems Conference*, Cambridge, MA, 2001. MIT Press. Cited on page(s) 71, 159

[168] Jon Kleinberg. The wireless epidemic. *Nature, Vol. 449*, Sep 2007. Cited on page(s) 123, 130

[169] Jon Kleinberg, Ravi Kumar, Prabhakar Raghavan, Sridhar Rajagopalan, and Andrew Tomkins. The web as a graph: Measurements, models and methods. In *International Computing and Combinatorics Conference*, Berlin, Germany, 1999. Springer. Cited on page(s) 10, 11, 47, 48, 58, 59, 60

[170] Tamara G. Kolda and Brett W. Bader. Tensor decompositions and applications. *SIAM Review*, 51(3):455–500, 2009. Cited on page(s) 108

[171] Tamara G. Kolda and Dianne P. O'Leary. A semidiscrete matrix decomposition for latent semantic indexing information retrieval. *ACM Trans. Inf. Syst.*, 16(4):322–346, 1998. Cited on page(s) 101

[172] Tamara G. Kolda and Jimeng Sun. Scalable tensor decompositions for multi-aspect data mining. In *ICDM*, pages 363–372. IEEE Computer Society, 2008. Cited on page(s) 108

[173] T.G. Kolda and B.W. Bader. The tophits model for higher-order web link analysis. In *Workshop on Link Analysis, Counterterrorism and Security*, volume 7, pages 26–29, 2006. Cited on page(s) 108

[174] T.G. Kolda and B.W. Bader. Tensor decompositions and applications. *SIAM review*, 51(3), 2009. Cited on page(s) 108

[175] Yehuda Koren, Stephen C. North, and Chris Volinsky. Measuring and extracting proximity in networks. In *KDD*, pages 245–255, 2006. Cited on page(s) 136

[176] Paul L. Krapivsky and Sidney Redner. Organization of growing random networks. *Physical Review E*, 63(6):066123 1–14, 2001. Cited on page(s) 55, 61, 62

[177] Valdis E. Krebs. Mapping networks of terrorist cells. *Connections*, 24(3):43–52, 2001. Cited on page(s) 153

[178] Ravi Kumar, Prabhakar Raghavan, Sridhar Rajagopalan, D. Sivakumar, Andrew Tomkins, and Eli Upfal. Stochastic models for the Web graph. In *IEEE Symposium on Foundations of Computer Science*, Los Alamitos, CA, 2000. IEEE Computer Society Press. Cited on page(s) 59

[179] Ravi Kumar, Prabhakar Raghavan, Sridhar Rajagopalan, and Andrew Tomkins. Extracting large-scale knowledge bases from the web. In *Proc. of Intl. Conf. on Very Large Data Bases*, San Francisco, CA, 1999. Morgan Kaufmann. Cited on page(s) 37, 47, 48, 58, 59, 60, 64, 118

[180] Michihiro Kuramochi and George Karypis. Frequent subgraph discovery. In *IEEE International Conference on Data Mining*, pages 313–320, Los Alamitos, CA, 2001. IEEE Computer Society Press. Cited on page(s) 156

[181] Michihiro Kuramochi and George Karypis. Discovering frequent geometric subgraphs. In *IEEE International Conference on Data Mining*, Los Alamitos, CA, 2002. IEEE Computer Society Press. Cited on page(s) 156

[182] Daniel D. Lee and H. Sebastian Seung. Algorithms for non-negative matrix factorization. In *NIPS*, pages 556–562, 2000. Cited on page(s) 101

[183] Jure Leskovec and Eric Horvitz. Planetary-scale views on a large instant-messaging network. In *WWW*, pages 915–924, 2008. Cited on page(s) 15, 163

[184] Jure Leskovec, Jon Kleinberg, and Christos Faloutsos. Graphs over time: densification laws, shrinking diameters and possible explanations. In *KDD '05: Proceeding of the eleventh ACM SIGKDD international conference on Knowledge discovery in data mining*, pages 177–187, New York, NY, USA, 2005. ACM Press. Cited on page(s) 4

[185] Jure Leskovec, Jon M. Kleinberg, and Christos Faloutsos. Graphs over time: densification laws, shrinking diameters and possible explanations. In *Proceedings of the Eleventh ACM SIGKDD International Conference on Knowledge Discovery and Data Mining, Chicago, Illinois, USA, August 21-24, 2005*, pages 177–187, 2005. Cited on page(s) 41

[186] Jure Leskovec, Jon M. Kleinberg, and Christos Faloutsos. Graphs over time: densification laws, shrinking diameters and possible explanations. In *KDD*, pages 177–187, 2005. Cited on page(s) 161

[187] Jure Leskovec, Kevin J. Lang, Anirban Dasgupta, and Michael W. Mahoney. Statistical properties of community structure in large social and information networks. In *WWW*, pages 695–704, 2008. Cited on page(s) 120

[188] Jure Leskovec, Mary Mcglohon, Christos Faloutsos, Natalie Glance, and Matthew Hurst. Cascading behavior in large blog graphs: Patterns and a model. In *Society of Applied and Industrial Mathematics: Data Mining (SDM07)*, 2007. Cited on page(s) 5

[189] Jure Leskovec, Mary McGlohon, Christos Faloutsos, Natalie S. Glance, and Matthew Hurst. Patterns of cascading behavior in large blog graphs. In *Proceedings of the Seventh SIAM International Conference on Data Mining, April 26-28, 2007, Minneapolis, Minnesota, USA*, 2007. Cited on page(s) 41

[190] Jurij Leskovec, Deepayan Chakrabarti, Jon Kleinberg, and Christos Faloutsos. Realistic, mathematically tractable graph generation and evolution, using Kronecker Multiplication. In *Conference on Principles and Practice of Knowledge Discovery in Databases*, Berlin, Germany, 2005. Springer. Cited on page(s) 65, 84

[191] Jurij Leskovec, Jon Kleinberg, and Christos Faloutsos. Graphs over time: Densification laws, shrinking diameters and possible explanations. In *Conference of the ACM Special Interest Group on Knowledge Discovery and Data Mining*, New York, NY, 2005. ACM Press. Cited on page(s) 19, 20, 47, 48, 55, 61, 64, 89

[192] David Liben-Nowell and Jon Kleinberg. The link prediction problem for social networks. In *Proc. CIKM*, 2003. Cited on page(s) 136

[193] Xia Lin, Dagobert Soergel, and Gary Marchionini. A self-organizing semantic map for information retrieval. In *Proc. of ACM SIGIR*, pages 262–269, Chicago, IL, Oct. 13-16 1991. Cited on page(s) 98

[194] Yang-Yu Liu, Jean-Jacques Slotine, and Albert-Laszlo Barabasi. Controllability of complex networks. *Nature*, 43:123–248, May 12 2011. Cited on page(s) 162

[195] Linyuan Lu. The diameter of random massive graphs. In *Proceedings of the twelfth annual ACM-SIAM symposium on Discrete algorithms*, SODA '01, pages 912–921, Philadelphia, PA, USA, 2001. Society for Industrial and Applied Mathematics. Cited on page(s) 21

[196] Sara C. Madeira and Arlindo L. Oliveira. Biclustering algorithms for biological data analysis: A survey. *IEEE Transactions on Computational Biology and Bioinformatics*, 1(1):24–45, 2004. Cited on page(s) 115

[197] Robert May. Simple mathematical models with very complicated dynamics. *Nature*, 261 459, 1976. Lessons of chaos in population biology. Cited on page(s) 163

[198] ROBERT M. MAY. Will a large complex system be stable? *Nature*, 238(5364):413–414, Aug 1972. Cited on page(s) 163

[199] M. McGlohon, J. Leskovec, C. Faloutsos, N. Glance, and M. Hurst. Finding patterns in blog shapes and blog evolution. In *International Conference on Weblogs and Social Media.*, Boulder, Colo., Helen Martin 2007. Cited on page(s) 41

[200] Mary McGlohon. Structural analysis of networks: Observations and applications. Ph.D. thesis CMU-ML-10-111, Machine Learning Department, Carnegie Mellon University, December 2010. Cited on page(s) 21, 24, 27, 28, 30

[201] Mary McGlohon, Leman Akoglu, and Christos Faloutsos. Weighted graphs and disconnected components: patterns and a generator. In *Proceedings of the 14th ACM SIGKDD International Conference on Knowledge Discovery and Data Mining, Las Vegas, Nevada, USA, August 24-27, 2008*, pages 524–532, 2008. Cited on page(s) 12, 21, 24, 41

[202] Mary Mcglohon, Leman Akoglu, and Christos Faloutsos. Weighted graphs and disconnected components: Patterns and a generator. In *ACM Special Interest Group on Knowledge Discovery and Data Mining (SIG-KDD)*, August 2008. Cited on page(s) 27, 28, 30

[203] Mary McGlohon, Stephen Bay, Markus G. Anderle, David M. Steier, and Christos Faloutsos. Snare: a link analytic system for graph labeling and risk detection. In *Proceedings of the 15th ACM SIGKDD International Conference on Knowledge Discovery and Data Mining, Paris, France, June 28 - July 1, 2009*, pages 1265–1274, 2009. Cited on page(s) 3

[204] Amy McGovern and David Jensen. Identifying predictive structures in relational data using multiple instance learning. In *International Conference on Machine Learning*, Menlo Park, CA, 2003. AAAI Press. Cited on page(s) 155

[205] Alberto Medina, Ibrahim Matta, and John Byers. On the origin of power laws in Internet topologies. In *Conference of the ACM Special Interest Group on Data Communications (SIG-COMM)*, pages 18–34, New York, NY, 2000. ACM Press. Cited on page(s) 47, 48, 70

[206] Milena Mihail and Christos H. Papadimitriou. On the eigenvalue power law. In *International Workshop on Randomization and Approximation Techniques in Computer Science*, Berlin, Germany, 2002. Springer Verlag. Cited on page(s) 12

[207] Stanley Milgram. The small-world problem. *Psychology Today*, 2:60–67, 1967. Cited on page(s) 158, 160

[208] R. Milo, S. Shen-Orr, S. Itzkovitz, N. Kashtan, D. Chklovshii, and U. Alon. Network motifs: Simple building blocks of complex networks. *Science*, 298:824–827, 2002. Cited on page(s) 156

[209] Einat Minkov and William W. Cohen. An email and meeting assistant using graph walks. In *CEAS*, 2006. Cited on page(s) 136

[210] Michael Mitzenmacher. Dynamic models for file sizes and double pareto distributions. *Internet Mathematics*, 1(3):305–333, 2003. Cited on page(s) 38

[211] Alan L. Montgomery and Christos Faloutsos. Identifying Web browsing trends and patterns. *IEEE Computer*, 34(7):94–95, 2001. Cited on page(s) 11, 83

[212] James Moody. Race, school integration, and friendship segregation in America. *American Journal of Sociology*, 107(3):679–716, 2001. Cited on page(s) 113

[213] M. E. J. Newman. Power laws, pareto distributions and zipf's law. *Contemporary Physics*, 46, 2005. Cited on page(s) 10, 27

[214] Mark E. J. Newman. The structure and function of complex networks. *SIAM Review*, 45:167–256, 2003. Cited on page(s) 15, 113

[215] Mark E. J. Newman. Power laws, pareto distributions and Zipf's law. *Contemporary Physics*, 46:323–351, 2005. Cited on page(s) 10, 35, 36, 37

[216] Mark E. J. Newman, Stephanie Forrest, and Justin Balthrop. Email networks and the spread of computer viruses. *Physical Review E*, 66(3):035101 1–4, 2002. Cited on page(s) 79

[217] Mark E. J. Newman, Michelle Girvan, and J. Doyne Farmer. Optimal design, robustness and risk aversion. *Physical Review Letters*, 89(2):028301 1–4, 2002. Cited on page(s) 73

[218] Mark E. J. Newman, Steven H. Strogatz, and Duncan J. Watts. Random graphs with arbitrary degree distributions and their applications. *Physical Review E*, 64(2):026118 1–17, 2001. Cited on page(s) 47, 48, 52

[219] Siegfried Nijssen and Joost Kok. Faster association rules for multiple relations. In *International Joint Conference on Artificial Intelligence*, San Francisco, CA, 2001. Morgan Kaufmann. Cited on page(s) 157

[220] Christopher Palmer, Phil B. Gibbons, and Christos Faloutsos. ANF: A fast and scalable tool for data mining in massive graphs. In *Conference of the ACM Special Interest Group on Knowledge Discovery and Data Mining*, New York, NY, 2002. ACM Press. Cited on page(s) 13, 14, 73, 77, 78, 130

[221] Christopher Palmer and J. Gregory Steffan. Generating network topologies that obey power laws. In *IEEE Global Telecommunications Conference*, Los Alamitos, CA, November 2000. IEEE Computer Society Press. Cited on page(s) 47, 48, 51

[222] Christopher R. Palmer and Christos Faloutsos. Electricity based external similarity of categorical attributes. *PAKDD 2003*, April-May 2003. Cited on page(s) 136

[223] Jia-Yu Pan, Hiroyuki Kitagawa, Christos Faloutsos, and Masafumi Hamamoto. Autosplit: Fast and scalable discovery of hidden variables in stream and multimedia databases. In *Proc. of the The Eighth Pacific-Asia Conference on Knowledge Discovery and Data Mining (PAKDD 2004)*, Sydney, Australia, May 26-28 2004. (Best Student Paper Award). Cited on page(s) 101

[224] Jia-Yu Pan, Hyung-Jeong Yang, Christos Faloutsos, and Pinar Duygulu. Automatic multi-media cross-modal correlation discovery. In *KDD*, pages 653–658, 2004. Cited on page(s) 136

[225] Jia-Yu Pan, Hyung-Jeong Yang, Christos Faloutsos, and Pinar Duygulu. Automatic multi-media cross-modal correlation discovery. *KDD*, pages 653–658, 2004. Cited on page(s) 137, 138, 139

[226] Shashank Pandit, Duen Horng Chau, Samuel Wang, and Christos Faloutsos. Netprobe: a fast and scalable system for fraud detection in online auction networks. In *Proceedings of the 16th International Conference on World Wide Web, WWW 2007, Banff, Alberta, Canada, May 8-12, 2007*, pages 201–210, 2007. Cited on page(s) 1

[227] Gopal Pandurangan, Prabhakar Raghavan, and Eli Upfal. Using PageRank to characterize Web structure. In *International Computing and Combinatorics Conference*, Berlin, Germany, 2002. Springer. Cited on page(s) 11, 47, 48, 63

[228] E.E. Papalexakis and N.D. Sidiropoulos. Co-clustering as multilinear decomposition with sparse latent factors. In *Acoustics, Speech and Signal Processing (ICASSP), 2011 IEEE International Conference on*, pages 2064–2067. IEEE, 2011. Cited on page(s) 108

[229] R. Pastor-Santorras and A. Vespignani. Epidemic spreading in scale-free networks. *Physical Review Letters 86*, 14, 2001. Cited on page(s) 123, 128, 130

[230] Romualdo Pastor-Satorras and Alessandro Vespignani. Immunization of complex networks. *Physical Review E*, 65(3):036104 1–8, 2002. Cited on page(s) 130

[231] David M. Pennock, Gary W. Flake, Steve Lawrence, Eric J. Glover, and C. Lee Giles. Winners don't take all: Characterizing the competition for links on the Web. *Proceedings of the National Academy of Sciences*, 99(8):5207–5211, 2002. Cited on page(s) 37, 47, 48, 65, 81, 82, 84

[232] B. Aditya Prakash, Deepayan Chakrabarti, Michalis Faloutsos, Nicholas Valler, and Christos Faloutsos. Got the flu (or mumps)? check the eigenvalue! *arXiv:1004.0060v1 [physics.soc-ph]*, 2010. Cited on page(s) 123, 125, 127, 131, 133

[233] B. Aditya Prakash, Hanghang Tong, Nicholas Valler, Michalis Faloutsos, and Christos Faloutsos. Virus propagation on time-varying networks: Theory and immunization algorithms. In José L. Balcázar, Francesco Bonchi, Aristides Gionis, and Michèle Sebag, editors, *ECML/PKDD (3)*, volume 6323 of *Lecture Notes in Computer Science*, pages 99–114. Springer, 2010. Cited on page(s) 131

[234] William H. Press, Brian P. Flannery, Saul A. Teukolsky, and William T. Vetterling. *Numerical Recipes in C*. Cambridge University Press, 1988. Cited on page(s) 97

[235] William H. Press, Saul A. Teukolsky, William T. Vetterling, and Brian P. Flannery. *Numerical Recipes in C*. Cambridge University Press, Cambridge, UK, 2nd edition, 1992. Cited on page(s) 84, 98, 101

[236] Erzsébet Ravasz and Albert-László Barabási. Hierarchical organization in complex networks. *Physical Review E*, 65:026112 1–7, September 2002. Cited on page(s) 115

[237] Sidney Redner. How popular is your paper? an empirical study of the citation distribution. *The European Physics Journal B*, 4:131–134, 1998. Cited on page(s) 11, 37

[238] W. Reed and M. Jorgensen. The double pareto-lognormal distribution - a new parametric model for size distribution. *Communications in Statistics -Theory and Methods*, 33(8):1733 – 1753, 2004. Cited on page(s) 39

[239] Matthew Richardson and Pedro Domingos. Mining knowledge-sharing sites for viral marketing. In *Conference of the ACM Special Interest Group on Knowledge Discovery and Data Mining*, pages 61–70, New York, NY, 2002. ACM Press. Cited on page(s) 83, 160, 162

[240] Helge Ritter, Thomas Martinetz, and Klaus Schulten. *Neural Computation and Self-Organizing Maps*. Addison Wesley, Reading, MA, 1992. Cited on page(s) 98

[241] Arnold L. Rosenberg and Lenwood S. Heath. *Graph Separators, with Applications*. Kluwer Academic/Plenum Pulishers, 2001. Cited on page(s) 120

[242] G. Salton, E.A. Fox, and H. Wu. Extended boolean information retrieval. *CACM*, 26(11):1022–1036, November 1983. Cited on page(s) 98

[243] Manfred Schroeder. *Fractals, Chaos, Power Laws: Minutes from an Infinite Paradise*. W.H. Freeman and Company, New York, 1991. Cited on page(s) 163

[244] Michael F. Schwartz and David C. M. Wood. Discovering shared interests using graph analysis. *Communications of the ACM*, 36(8):78–89, 1993. Cited on page(s) 113

[245] Mukund Seshadri, Sridhar Machiraju, Ashwin Sridharan, Jean Bolot, Christos Faloutsos, and Jure Leskovec. Mobile call graphs: beyond power-law and lognormal distributions. In *Proceedings of the 14th ACM SIGKDD International Conference on Knowledge Discovery and Data Mining, Las Vegas, Nevada, USA, August 24-27, 2008*, pages 596–604, 2008. Cited on page(s) 38

[246] Jianbo Shi and Jitendra Malik. Normalized cuts and image segmentation. *IEEE Trans. on Pattern Analysis and Machine Intelligence*, 22(8):888–905, 2000. Cited on page(s) 137

[247] G. Siganos, M. Faloutsos, P. Faloutsos, and C. Faloutsos. Power laws and the AS-level internet topology, 2003. Cited on page(s) 11, 41

[248] G. Siganos, S. L. Tauro, and M. Faloutsos. Jellyfish: a conceptual model for the as internet topology. *Journal of Communications and Networks*, 2006. Cited on page(s) 13

[249] Herbert Simon. On a class of skew distribution functions. *Biometrika*, 42(3/4):425–440, 1955. Cited on page(s) 53

[250] R. V. Solé and J. M. Montoya. Complexity and fragility in ecological networks. In *Proceedings of the Royal Society of London B*, volume 268, pages 2039–2045, London, UK, 2001. The Royal Society. Cited on page(s) 79

[251] Malcolm K. Sparrow. The application of network analysis to criminal intelligence: An assessment of the prospects. *Social Networks*, 13(3):251–274, 1991. Cited on page(s) 153

[252] Daniel A. Spielman and Shang-Hua Teng. Spectral partitioning works: Planar graphs and finite element meshes. In *IEEE Symposium on Foundations of Computer Science*, pages 96–105, Los Alamitos, CA, 1996. IEEE Computer Society Press. Cited on page(s) 117

[253] Shlomo Sternberg. *Dynamical Systems*. Dover Publications Inc., 2010. Cited on page(s) 97, 104

[254] Gilbert Strang. *Linear Algebra and Its Applications*. Academic Press, 2nd edition, 1980. Cited on page(s) 97, 101

[255] Jimeng Sun, Huiming Qu, Deepayan Chakrabarti, and Christos Faloutsos. Neighborhood formation and anomaly detection in bipartite graphs. In *ICDM*, pages 418–425, 2005. Cited on page(s) 136

[256] Jimeng Sun, Charalampos E. Tsourakakis, Evan Hoke, Christos Faloutsos, and Tina Eliassi-Rad. Two heads better than one: pattern discovery in time-evolving multi-aspect data. *Data Min. Knowl. Discov.*, 17(1):111–128, 2008. Cited on page(s) 108

[257] J.T. Sun, H.J. Zeng, H. Liu, Y. Lu, and Z. Chen. Cubesvd: a novel approach to personalized web search. In *Proceedings of the 14th international conference on World Wide Web*, pages 382–390. ACM, 2005. Cited on page(s) 108

[258] Nassim Nicholas Taleb. *The Black Swan: The Impact of the Highly Improbable*. Random House, 2007. Cited on page(s) 10, 11

[259] Hongsuda Tangmunarunkit, Ramesh Govindan, Sugih Jamin, Scott Shenker, and Walter Willinger. Network topologies, power laws, and hierarchy. Technical Report 01-746, University of Southern California, 2001. Cited on page(s) 13, 14, 75, 77, 78

[260] Hongsuda Tangmunarunkit, Ramesh Govindan, Sugih Jamin, Scott Shenker, and Walter Willinger. Network topology generators: Degree-based vs. structural. In *Conference of the ACM Special Interest Group on Data Communications (SIGCOMM)*, New York, NY, 2002. ACM Press. Cited on page(s) 77, 78

[261] Sudhir L. Tauro, Christopher Palmer, Georgos Siganos, and Michalis Faloutsos. A simple conceptual model for the Internet topology. In *Global Internet*, Los Alamitos, CA, 2001. IEEE Computer Society Press. Cited on page(s) 14, 31, 78

[262] Jos M. F. ten Berge. *Least squares optimization in multivariate analysis*. DSWO Press, Leiden University, 1993. ISBN-10: 9066950838, ISBN-13: 978-9066950832, http://www.rug.nl/psy/onderzoek/onderzoeksprogrammas/pdf/leastsquaresbook.pdf. Cited on page(s) 97, 100

[263] Robert Tibshirani, Guenther Walther, and Trevor Hastie. Estimating the number of clusters in a dataset via the Gap statistic. *Journal of the Royal Statistical Society, B*, 63:411–423, 2001. Cited on page(s) 117

[264] Hanghang Tong. *Fast Algorithms for Querying and Mining Large Graphs*. PhD thesis, CMU, September 2009. Cited on page(s) 130, 136

[265] Hanghang Tong and Christos Faloutsos. Center-piece subgraphs: problem definition and fast solutions. In *KDD*, pages 404–413, 2006. Cited on page(s) 136

[266] Hanghang Tong and Christos Faloutsos. Center-piece subgraphs: problem definition and fast solutions. In *Proceedings of the Twelfth ACM SIGKDD International Conference on Knowledge Discovery and Data Mining, Philadelphia, PA, USA, August 20-23, 2006*, pages 404–413, 2006. Cited on page(s) 141, 143

[267] Hanghang Tong, B. Aditya Prakash, Charalampos E. Tsourakakis, Tina Eliassi-Rad, Christos Faloutsos, and Duen Horng Chau. On the vulnerability of large graphs. In Geoffrey I. Webb, Bing Liu, Chengqi Zhang, Dimitrios Gunopulos, and Xindong Wu, editors, *ICDM*, pages 1091–1096. IEEE Computer Society, 2010. Cited on page(s) 130

[268] Jeffrey Travers and Stanley Milgram. An experimental study of the Small World problem. *Sociometry*, 32(4):425–443, 1969. Cited on page(s) 13, 41, 67, 157, 158, 160

[269] Charalampos Tsourakakis. Counting of triangles in large real networks, without counting: Algorithms and laws. *ICDM*, 2008. Cited on page(s) 41

[270] Charalampos E. Tsourakakis. Fast counting of triangles in large real networks without counting: Algorithms and laws. In *ICDM*, 2008. Cited on page(s) 15

[271] Joshua R. Tyler, Dennis M. Wilkinson, and Bernardo A. Huberman. *Email as spectroscopy: Automated discovery of community structure within organizations*, pages 81–96. Kluwer, B.V., Deventer, The Netherlands, 2003. Cited on page(s) 116

[272] Nicholas Valler, B. Aditya Prakash, Hanghang Tong, Michalis Faloutsos, and Christos Faloutsos. Epidemic spread in mobile ad hoc networks: Determining the tipping point. In Jordi

Domingo-Pascual, Pietro Manzoni, Sergio Palazzo, Ana Pont, and Caterina M. Scoglio, editors, *Networking (1)*, volume 6640 of *Lecture Notes in Computer Science*, pages 266–280. Springer, 2011. Cited on page(s) 131

[273] Stijn Marinus van Dongen. *Graph clustering by flow simulation*. PhD thesis, Univesity of Utrecht, 2000. Cited on page(s) 117

[274] M. Vasilescu and D. Terzopoulos. Multilinear analysis of image ensembles: Tensorfaces. *Computer Vision ECCV 2002*, pages 447–460, 2002. Cited on page(s) 108

[275] Satu Virtanen. Clustering the Chilean Web. In *Latin American Web Congress*, Los Alamitos, CA, 2003. IEEE Computer Society Press. Cited on page(s) 118

[276] Yang Wang, Deepayan Chakrabarti, Chenxi Wang, and Christos Faloutsos. Epidemic spreading in real networks: An eigenvalue viewpoint. In *Symposium on Reliable Distributed Systems*, pages 25–34, Los Alamitos, CA, 2003. IEEE Computer Society Press. Cited on page(s) 123

[277] Stanley Wasserman and Katherine Faust. *Social Network Analysis: Methods and Applications*. Cambridge University Press, Cambridge, UK, 1994. Cited on page(s) 147, 162

[278] Duncan J. Watts, Peter Sheridan Dodds, and Mark E. J. Newman. Identity and search in social networks. *Science*, 296:1302–1305, 2002. Cited on page(s) 71, 159

[279] Duncan J. Watts and Steven H. Strogatz. Collective dynamics of 'small-world' networks. *Nature*, 393:440–442, 1998. Cited on page(s) 15, 47, 48, 68, 113, 114, 115, 153, 158

[280] Bernard M. Waxman. Routing of multipoint connections. *IEEE Journal on Selected Areas in Communications*, 6(9):1617–1622, December 1988. Cited on page(s) 47, 48, 69

[281] Margaret R. Weeks, Scott Clair, Steve Borgatti, Kim Radda, and Jean J. Schensul. Social networks of drug users in high-risk sites: Finding the connections. *AIDS and Behavior*, 6(2):193–206, 2002. Cited on page(s) 147

[282] Virginia Vassilevska Williams. Breaking the coppersmith-winograd barrier, 2012. Cited on page(s) 14

[283] Jared Winick and Sugih Jamin. Inet-3.0: Internet Topology Generator. Technical Report CSE-TR-456-02, University of Michigan, Ann Arbor, 2002. Cited on page(s) 47, 48, 76, 77, 78

[284] Fang Wu and Bernardo A. Huberman. Finding communities in linear time: a physics approach. *The European Physics Journal B*, 38(2):331–338, 2004. Cited on page(s) 119

[285] Xifeng Yan and Jiawei Han. gSpan: Graph-based substructure pattern mining. In *IEEE International Conference on Data Mining*, Los Alamitos, CA, 2002. IEEE Computer Society Press. Cited on page(s) 156

190 BIBLIOGRAPHY

[286] Soon-Hyung Yook, Hawoong Jeong, and Albert-László Barabási. Modeling the Internet's large-scale topology. *Proceedings of the National Academy of Sciences*, 99(21):13382–13386, 2002. Cited on page(s) 47, 48, 71

[287] Stephen J. Young and Edward R. Scheinerman. Random dot product graph models for social networks. In *WAW*, pages 138–149, 2007. Cited on page(s) 89

[288] Philip S. Yu, Jiawei Han, and Christos Faloutsos. *Link Mining: Models, Algorithms, and Applications*. Springer, 2010. ISBN-10: 1441965149 ISBN-13: 978-1441965141. Cited on page(s) 162

[289] G.K. Zipf. *Human Behavior and Principle of Least Effort: An Introduction to Human Ecology*. Addison Wesley, Cambridge, Massachusetts, 1949. Cited on page(s) 9

Authors' Biographies

DEEPAYAN CHAKRABARTI

Dr. Deepayan Chakrabarti obtained his Ph.D. from Carnegie Mellon University in 2005. He was a Senior Research Scientist with Yahoo, and now with Facebook Inc. He has published over 35 refereed articles and is the co-inventor of the *RMat* graph generator (the basis of the *graph500* supercomputer benchmark). He is the co-inventor in over 20 patents (issued or pending). He has given tutorials in CIKM and KDD, and his interests include graph mining, computational advertising, and web search.

CHRISTOS FALOUTSOS

Christos Faloutsos is a Professor at Carnegie Mellon University and an ACM Fellow. He has received the Research Contributions Award in ICDM 2006, the SIGKDD Innovations Award (2010), 18 "best paper" awards (including two "test of time" awards), and four teaching awards. He has published over 200 refereed articles, and has given over 30 tutorials. His research interests include data mining for graphs and streams, fractals, and self-similarity, database performance, and indexing for multimedia and bio-informatics data.

Printed in the United States
by Baker & Taylor Publisher Services